INSECT TRANSFORMATION

昆蟲的華麗變身 演化適應之路

AN EVOLUTIONARY ADAPTATION PATHWAY

編著者／ 葉文斌、楊曼妙、路光暉
Editors Wen-Bin Yeh、Man-Miao Yang、Kuang-Hui Lu

英文校稿／ 施劍鎣、梁國汶
English Revisers Chain-Ing Shih、Kok-Boon Neoh

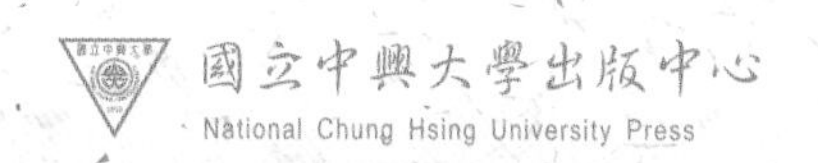

國立中興大學出版中心
National Chung Hsing University Press

序

PREFACE

經濟水平提高，文化素養增進後，人類社會對環境開發破壞的省思，也愈來愈強烈；保護自然、愛護大地的思緒及生態旅遊的活動受到很大的重視。佔動物種類百分之八十以上的昆蟲，具有極高的多樣性與普遍性，是推廣科學教育的好素材；本校的昆蟲標本蒐藏豐富，擁有相當多獨特，且具有歷史意義和研究重要性的標本，除提供教學研究用，也不定期舉辦昆蟲展示活動，進行環境教育的推廣。

昆蟲學系近十年來，在學校的支持下，舉辦多次的昆蟲展示活動。本人接任校長後，持續支持昆蟲相關的策展，設計各類具有教育意義的展示主題及推廣活動，達到環境教育與自然保育推廣目的：例如 2018 年的「招蜂引蝶生態花園展」，介紹蜜蜂及蝴蝶於植物授粉的重要性，更於 2019 年中興大學百年校慶時推出「興潮蝶起．BetterFly」為期一年的蝴蝶特展。很高興得知，昆蟲學系將目前環境教育常設展的部分內容，透過深入淺出的文字，輔以插圖、生態照片、油畫、水彩畫，按各個展示主題編輯成冊，並以中英文並列方式介紹，除進一步推廣環境保護議題外，也讓無法前來中興大學觀展的民眾可以一飽眼福。

長年以來，中興大學在教學與研究上穩健紮根與蓬勃發展，同時也重視社會教育與科學知識的推廣。近年學校更積極籌劃校史館的設立，除介紹中興大學過去在農業及自然科學上的傑出表現外，也介紹人文、歷史及社會科學各領域的重要貢獻；為發揮校史館最大功能，館內將預留空間，讓學校各單位策劃展覽，進行科普教育；相信包含昆蟲學系在內的各單位，都能透過展題的設計，傳達人類文明與生物演化的知識和理念，進而達到推廣自然科學及人文社會教育的目標，展現中興大學的社會責任及歷史使命。

With the economic growth and enrichment of cultural competence, human beings started to introspect their past activities about environmental destruction. The thought of protecting the nature, taking care of our mother earth, and eco-tourism activities have received increasing attention. Insects, which account for more than 80% of animal species, are extremely diverse and widespread, and are good materials for promoting science education. Entomological Museum in National Chung Hsing University has a rich and unique collection of insect specimens which showed its significance in nature history and research importance. These insect specimens are used not only for teaching and research purposes but also for insect exhibitions to promote environmental education.

The Department of Entomology has organized many insect exhibitions with the supports of the university for the past ten years. After I was appointed as the president, I continued to support and encourage the department to hold insect exhibitions on a routine basis and to design educational activities that related to insect in effort to promote the environmental education and nature conservation. For example, in 2018, the "Exhibition of Bees and Butterflies: Ecological Garden" introduced the importance of bees and butterflies in pollination of plants. In 2019, in celebrating the centennial anniversary of the university, the department held an one-year butterfly exhibition — "Butterflies: Betterfly." I am pleased to see that all of the faculty members in Department of Entomology devote themselves to compile and edit the contents of "the exhibition of environmental education" into a book that is bilingual in Mandarin and English. The contents of the book are supplemented by insect illustrations, photos, oil paintings, and watercolors. The book allows people who are unable to visit the exhibition have the chance to enjoy, not to mention to promote environmental protection campaigns to the public.

Over the years, National Chung Hsing University has progressed and made a great deal of achievements in teaching and researches. Also, we highly concern about promoting social and scientific education. In recent years, we have actively planned to establish a history museum which is to emphasize the outstanding performance in agriculture and natural sciences, and to display the significant contributions in the fields of humanities, history, and social science. Moreover, in order to maximize the function of the university history museum, rooms in the museum will be reserved for all units and divisions of the university to hold the exhibitions and scientific education activities. I truly believed that through such activities, all units or divisions, including the Department of Entomology, would convey the ideas and knowledge of human civilization, natural and biological science to the public. All these efforts will help National Chung Hsing University to fulfill the social responsibilities and historical missions.

國立中興大學 校長
President, National Chung Hsing University

薛富盛 F.S. Shieu

序

PREFACE

在生物多樣性及生態保育的潮流下，昆蟲學者及愛好者陸續出版不少圖鑑介紹昆蟲，成為中小學重要的教學材料，也促使以昆蟲為題材的環境教育展示活動逐漸在國內開展。國立中興大學昆蟲學系在因緣際會下，獲得日本名古屋女子大學八田耕吉教授餽贈的數萬件珍貴昆蟲標本，在國內環境教育展示有成的台北科學教育館、台中科博館、台北市動物園等幾個單位及校方的協助支持下，於 2012 年在中興大學圖書館舉辦了「有蟲自遠方來」特展，獲得昆蟲學系系友、昆言企業股份有限公司黃吉金董事長的慷慨捐贈及學校同仁的重視，並接續展開了一系列專業的昆蟲展示活動，本人有幸在該期間擔任校長，躬逢其盛！於 2017 年昆蟲學系更進一步地推出了「昆蟲展示與環境教育」的常設展，充分利用昆蟲學系的公共空間，結合昆蟲學系原有的昆蟲標本展示室，充分且有效率地向社區民眾、中小學師生推廣昆蟲學知識，並有系統地介紹環境保育議題。

此一書籍的出版即是以上述「昆蟲展示與環境教育」的展示內容為主，介紹台灣對保育的付出及昆蟲的多樣特性。令人高興的是，昆蟲學系把此展覽內容編排成書出版，更進一步以中、英文並列的方式呈現，不僅讓華文世界的讀者得以快速吸收新知，了解「昆蟲世界」的樣貌，也讓精彩的昆蟲學知識及環境教育的內容得以介紹給全世界，加速國際化。

看到此書的出版，感到非常欣慰，之前播下的種子，逐漸成長、開花結果！也衷心地祝福昆蟲學系在環境推廣教育上持續貢獻，更期待創校已逾百年的國立中興大學在各領域的科普教育工作上都能有精彩傑出的表現。

Entomologists and enthusiasts have successively published many insect guide books to introduce insects. These books have become essential learning materials for students in elementary and high schools and are constantly used in insect-themed exhibitions to promote the environmental education in Taiwan. By serendipity, the Department of Entomology, National Chung Hsing University, has received tens of thousands of precious insect specimens as a donation from Professor Kokichi Hatta, Nagoya Women's University, Nagoya, Japan. In 2012, with the assistance and support of the National Taiwan Science Education Center, National Museum of Natural Science, Taipei Zoo, etc., the Department of Entomology organized a special exhibition — "Insect Exhibition from Far Away", which was held in the hall of the University Library. With a generous donation from Chairman Ji-Jin Huang of Entomek Enterprise Corp., an alumnus of the Department of Entomology, and the support of faculty and staff of the university, a series of professional insect exhibitions were thereafter organized and held. I had the honor to witness these events during my tenure as the President of the University. I am also very glad to learn that the Department of Entomology established in 2017 a permanent exhibition — "Insect Exhibition of Environmental Education" on campus, using the hallways and specimen exhibition room in the Department of Entomology building as the exhibition venue. The Exhibition will aim to provide inspiration and motivation of learning about insect, and to introduce environmental conservation issues and basic knowledge of entomology to the community, elementary and high school students and teachers.

The publication of this book is based on the contents of the aforementioned "Insect Exhibition of Environmental Education" which summarizes the achievements of conservation effort and the biodiversity of insects in Taiwan. I'd like to express my sincere gratitude for the effort of the colleagues in the Department of Entomology to compile and edit the book into a bilingual version in both Chinese and English. This book allows Chinese readers to readily absorb the knowledge about "the Insect World", and also introduces the international readers to insects and environmental education in Taiwan.

I am very pleased to see this book in print, and witness the seeds of knowledge planted earlier to get growing and blooming! With my wholehearted blessing I am hopeful that all the members in the Department of Entomology will continue devoting themselves to environmental conservation education, and moreover I am looking forward to seeing our centennial university make more splendid achievements in science education and popularization of science in various fields.

中央研究院 院士｜國立中興大學 前任校長
Academican, Academia Sinica, Taiwan
Former President, National Chung Hsing University

李德財 De-Tsai Lee

序

PREFACE

昆蟲在 4 億年的演化長河下，雖歷經劇烈地殼變動及環境氣候變遷，然沒因此而滅絕，反倒是開枝拓葉、源源不絕地演化出繁複多樣的種類，成為地球上最為成功的生物。昆蟲上天、下地、鑽土、入水，適應地球上的各種環境與棲所；牠化身為亮麗優雅的蝴蝶、看起來強壯威武的甲蟲，或是令人討厭的蒼蠅、蚊子、蟑螂、螞蟻等。總是昆蟲不僅是大自然的瑰寶，有許多更是與人類的食、衣、住、行、育、樂息息相關，正是環境科學教育最好的素材。

中興大學昆蟲學系長年以來每隔一年都會舉辦一次為期一週的昆蟲展覽，可以說是提供大台中地區中小學生自然科普教學的盛事；三十餘年來每次展覽都可以吸引 1~2 萬的人潮觀展，想必應該也灑播下無數的種子，讓環境科學教育向下紮根。而除了每兩年一次的短期展覽活動外，昆蟲學系亦會不定期舉辦規模不等的展覽。如 2010 年獲日本學者捐贈數萬件珍貴標本時，在校方的支持下舉辦了「有蟲自遠方來特展」。又如與國立臺灣科學教育館、國立自然科學博物館、台北市立動物園等單位合作展出 10 餘場昆蟲展覽。此外，亦會到中部地區各中小學、甚至幼兒園，進行以昆蟲為主題的環境教育推廣活動。

2017 年感謝李德財校長、呂福興教務長、等人的大力支持，協助昆蟲系設立「昆蟲環境教育展示」常設展，首創將昆蟲學系各樓層的中庭、走廊、樓梯間等空間化為展場，並以接受昆蟲愛好者及各級學校預約導覽的方式，有系統的推廣自然科學環境教育。此昆蟲環境教育展示，基本介紹昆蟲特性、各類特化的行為，並設計一系列演化適應的主題，透過插圖、生態照片與藝術創作等方式讓大眾了解生態資源及生物多樣性的重要，並得以思考如何永續應用地球上的珍貴資源、減少環境破壞，並與大自然環境和諧共存。今天為分享這些昆蟲科普知識予更多人，進一步將展示內容編輯成冊，並以中英文呈現，以期讓更多人知道昆蟲在環境教育上的重要性。

人類的經濟開發已停不下來，全球暖化所帶來的環境驟變正在形塑一個全然不同的生態環境，許多自然棲地逐漸喪失，嚴重威脅昆蟲與許多動植物的生存。期望透過這本書中的內容，喚醒我們珍惜大自然的意識，共同為保護大自然貢獻心力。

Under 400 million years of evolution, insects have gone through violent tectonic movement and environmental changes. Instead of extinction, they expanded their territory and evolved multiple times into diverse species which have made themselves the most successful creatures on the earth. Insects are ubiquitous — they can fly, craw, dive, and hide in the soil adapt to all kinds of environments and habitats. They transform themselves into the bright and fascinating butterflies, the strong and mighty beetles, or the annoying flies, mosquitoes, cockroaches, or ants, and others. Insects are not only the important treasures of the nature, but also vital for humane society in providing food, clothing, shelter, transportation, education, and entertainment. They are the best materials for environmental science studies.

For the past 30 years, the Department of Entomology, National Chung Hsing University has held a one-week insect exhibition biannually. The exhibition has been a grand educational activity for students in Taichung. There are about 10,000 to 20,000 people visit to the exhibition. The event immensely increased the awareness of environmental issues. In addition to organizing the exhibitions, the Department of Entomology also hold exhibitions of various scales from time to time. In 2010, tens of thousands of precious specimens have been donated by Japanese scholars to exhibit in the exhibition themed as "Insect Exhibition from Far Away". Collaborating with the National Taiwan Science Education Center, National Museum and Natural Science, Taipei Zoo, and other organizations, the Department has organized more than 10 insect exhibitions in the past 10 years. Moreover, we visited elementary and high schools, and kindergartens to carry out the environmental education that related with the theme of insect.

We thank President Der-Tsai Lee, Dean Fu-Hsing Lu, and others who had strongly supported the Department of Entomology to set up the permanent exhibition — "Insect Exhibition of Environmental Education." The room and walls in atrium, corridor, and other places on each floor of the Department of Entomology were used as the exhibiting places. Reservation for exhibition guide tour is open for insect enthusiasts and schools. The contents of exhibition focus on the information about insect such as insect characteristics and insect behaviors and a series of evolutionary adaptation. Through reading the illustrations, photos, paintings and artistic creation, visitors can understand the significance of ecological resources and biodiversity. The exhibition allows visitors to think about how to utilize the precious resources of the earth sustainably, how to reduce environmental risks effectively, and how to coexist with other species in nature harmoniously. This book is produced to share the knowledge of insect science and the significance of insects in both environment and education.

An unstoppable economic exploitation has caused dramatic climate changes. The changes are now shaping the environment today which has led to many natural habitats lost and seriously threatening the survival of insects, animals and plants. We sincerely hope that this book will raise public awareness of environmental protection.

國立中興大學昆蟲學系 前任系主任
Former Chairman of Department of Entomology,
National Chung Hsing University

序

PREFACE

「昆蟲的華麗變身：演化適應之路」一書的出版是結合不同領導階層與工作團隊共同努力的結晶，包括：薛富盛校長與李德財前校長的大力支持、杜武俊教授與路光暉教授兩位前系主任的辛勤耕耘、葉文斌教授與楊曼妙教授兩位系統分類與形態專家的指導，以及系上無數熱愛昆蟲的學生與系友們的投入。

「前人種樹後人乘涼」，因為你們的付出，讓我們有機會悠游其中、享受大自然的奧妙。惟願系上後起之秀，持續造福讀者，薪火相傳、一本接一本，讓我們逐一窺探「美妙的昆蟲世界」。

The publication of "Insect transformation: An evolutionary adaptation pathway" has accumulated tremendous efforts from different leaderships and work teams, including the strong supports from current president (Fuh-Sheng Shieu) and former president (Der-Tsai Lee) of National Chung Hsing University, the project development by two former department heads (Professors Wu-Chun Tu and Kuang-Hui Lu), the guidance of two assiduous and unremitting instructors (Professors Wen-Bin Yeh and Man-Miao Yang), and involvement and input of countless insect-devoting students in the Department of Entomology.

"The predecessors (forerunners) planted trees and the descendants (successors) enjoy the shade", because of your contributions, we have the opportunity to wander around and enjoy the mysteries of nature. I hope the remarkable younger generation will continue to benefit the readers, book by book, let us explore the "wonderful insect world" one by one.

國立中興大學昆蟲學系 系主任
Chairman of Department of Entomology,
National Chung Hsing University

目錄 CONTENT

你變態啊！昆蟲！

Insects! Metamorphosis!

吳奕徵、葉文斌
I-Chaen Wu、Wen-Bin Yeh

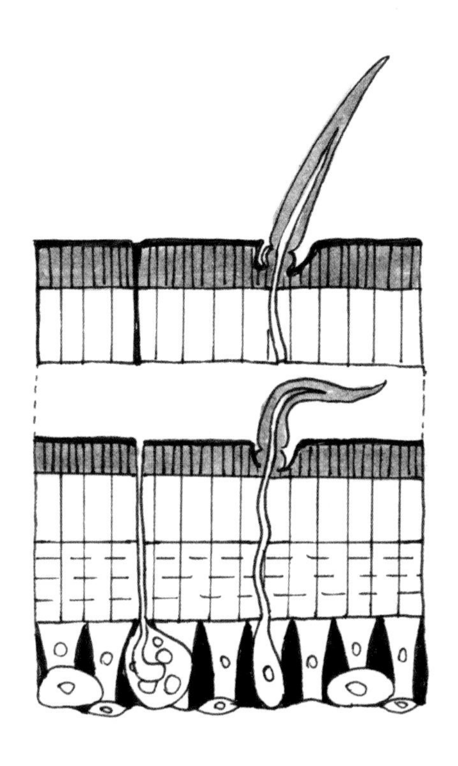
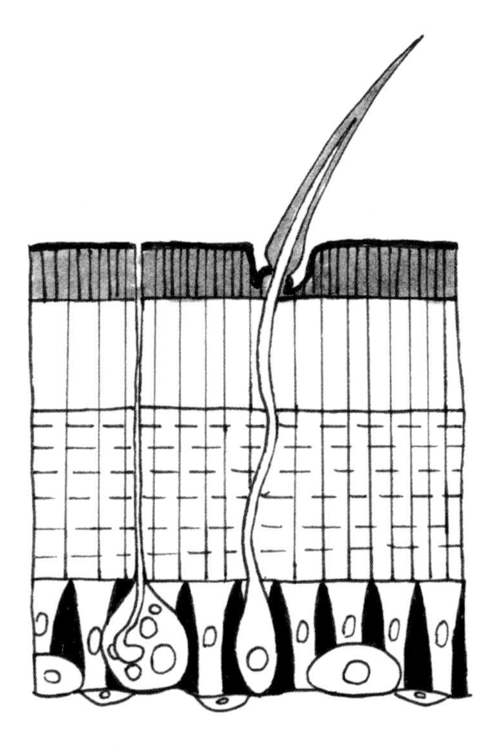

昆蟲舊表皮分解後，部分物質再吸收進入新表。
After partially breaking down, the old cuticle retracts and become integrated into the new cuticle.

一、你變態啊！昆蟲！

昆蟲為節肢動物門最大的一個綱，體型大小受幾丁質組成的外骨骼限制；要長大僅能經由蛻皮方式移除舊表皮，再合成較大的新表皮，遂而演化出不一樣的變態過程。

Insects! Metamorphosis!

Insecta is the largest class within the arthropod phylum. The body size of insects is restricted by their chitinous exoskeleton. Still, insects can grow by shedding and replacing old exoskeletons with new and larger ones, a process known as molting. Over time, insects have evolved various forms of metamorphosis.

(一)無變態：以不變應萬變

衣魚及石蛃等昆蟲，成長過程外型變化不大，幼期除了小、無生殖能力外，食性、習性、生態皆與成蟲似，為無變態昆蟲，幼蟲常以仔蟲 (Young) 稱之，分卵、『仔仔』與成蟲三階段。

Ametabolous: Keep calm than agitation

Silverfish and bristletail are so-called ametabolous insects because they undergo minimal or no metamorphosis. In ametabolous insects, juveniles are no different in appearance, feeding habit, and behavior from adults. Their life cycle includes egg, juvenile, and adult phases. Juveniles are simply referred to as "young." The only difference is that juveniles are smaller and sexually immature.

衣魚
Silverfish

石蛃
Bristletail

1. 繪 / 郭致與 (Sketch：Chih-Yu Kuo)
2. 圖 / 詹明澍 (Photo：Ming-Shu Chan)
3. 圖 / 江允中 (Photo：Yun-Chung Chiang)

1 | 2
  | 3

(二)不完全變態：一暝大一吋

Incomplete metamorphosis: Sleeping beauty

不完全變態昆蟲有兩類－「漸進變態」和「半行變態」，生長過程有卵、幼生期、成蟲三個階段。

Incomplete metamorphosis includes gradual- and hemi-metamorphosis. Insects that undergo incomplete metamorphosis have three life stages: egg, juvenile, and adult.

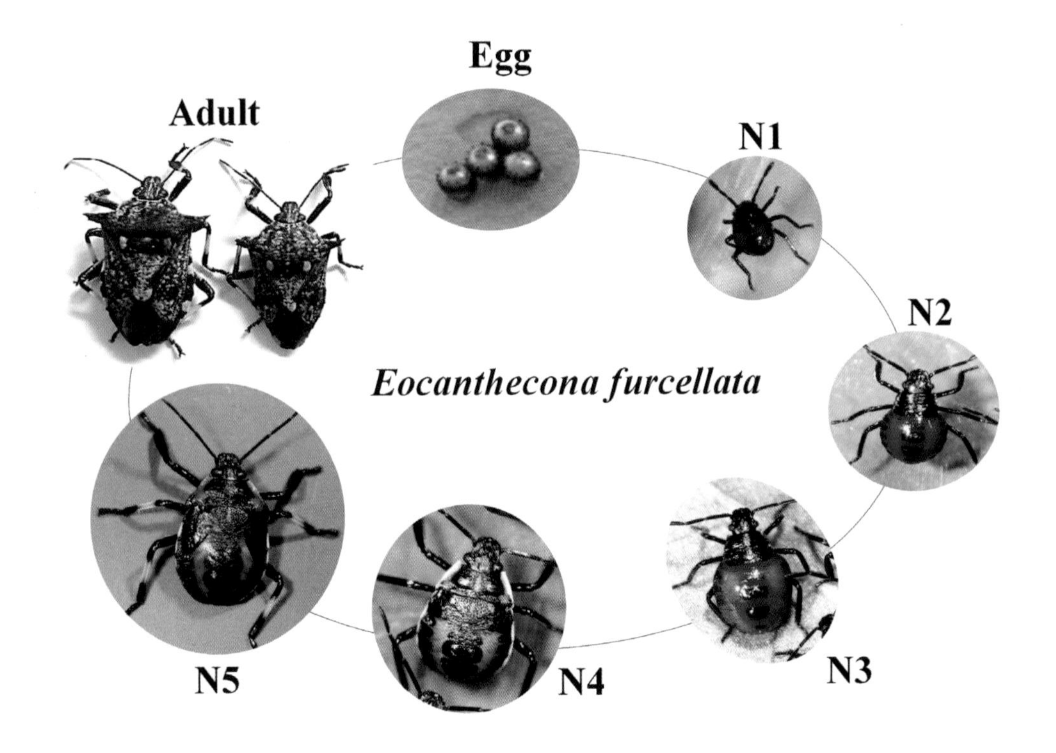

圖 / 段淑人（Photo：Shu-Jen Tuan）

黃斑粗喙椿象三個階段的生活史，即卵期、若蟲期及成蟲期；若蟲有 N1~N5 等 5 個齡期。
The true bug *Eocanthecona furcellata* undergoes three life stages: egg, nymph, and adult. Its nymphal stages are further divided into 5 instars.

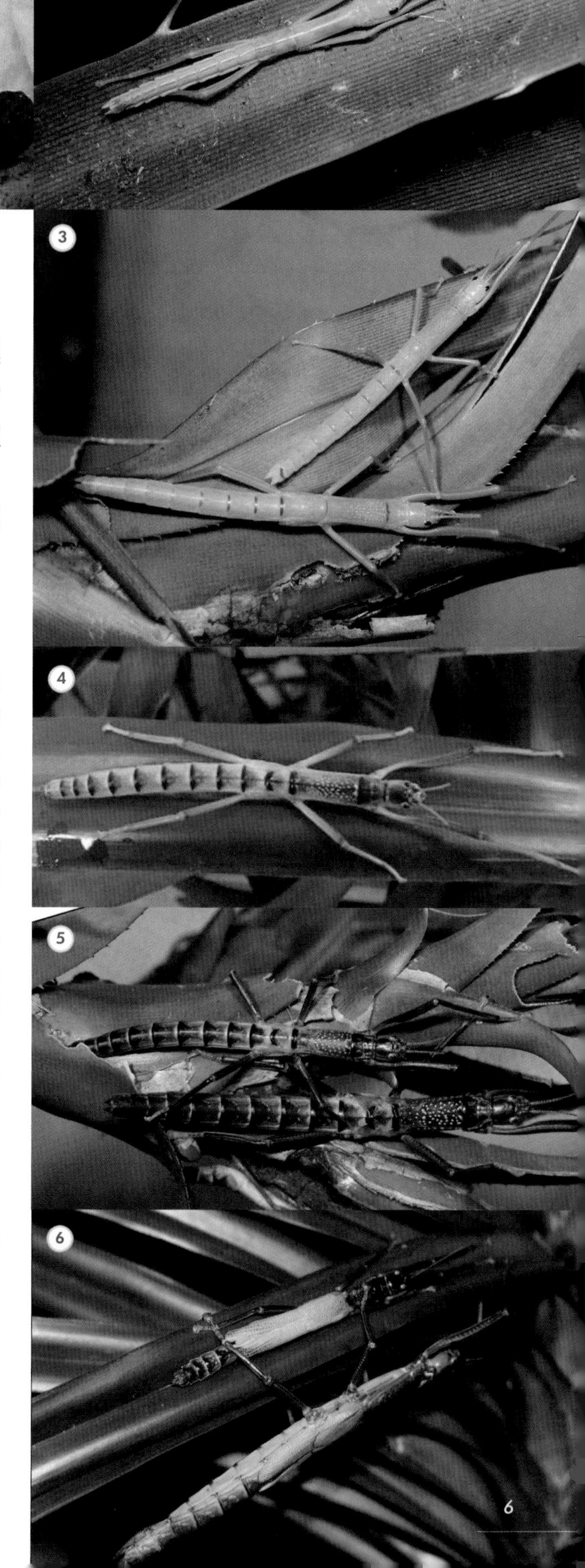

1. 漸進變態

漸進變態昆蟲成長過程外型變異不大，幼期與成蟲外型似，環境棲所也雷同，稱若蟲 (nymph)，翅芽累次脫皮漸長，如椿象、竹節蟲、蟑螂、蝗蟲等。牠們與半行變態最主要的區別就是若蟲與稚蟲。

Gradual metamorphosis

Insects that undergo gradual metamorphosis do not experience drastic changes from one life stage to another. In fact, the nymphs and adults of this group of insects share many similarities in their appearance and habitat use. The wing discs of nymphs grow at a gradual pace between molting events. True bugs, stick insects, cockroaches, and locusts are included in this group of insects. The main difference between hemi- and gradual metamorphosis is their juvenile stages, which are called nymphs and naiads, respectively.

圖 / 吳怡欣（Photo：I-Hsin Wu）
圖 / 游依靜（Photo：Yi-Ching Yu）
圖 / 蕭忠義（Photo：Chung-Yi Hsiao）

津田氏大頭竹節蟲生活史含卵、若蟲 6 個齡期及體型不同的雌雄成蟲。參考自 Wu et al. 2020
The life cycle of Tsuda's big-head stick insect, *Megacrania tsudai*, includes egg six nymphal stages, and the distinct adult male and female in their body size.（Cited from Wu et al. 2020）

2. 半行變態

半行變態的代表為石蠅、蜻蜓、豆娘及蜉蝣，幼生期被稱為稚蟲 (naiad)，成長過程變化與漸進變態昆蟲相似，但羽化成蟲後，環境棲所完全不同。

Hemi-metamorphosis

Stoneflies, dragonflies, damselflies, and mayflies are all examples of hemi-metabolous insects. Their juveniles are called "naiad", and live under water. In contrast, adults emerge above water where they complete their life cycle.

圖 / 詹明澍（Photo：Ming-Shu Chan）

白痣珈蟌成蟲
Adult damselfly, *Matrona cyanoptera*

圖 / 詹明澍（Photo：Ming-Shu Chan）

豆娘稚蟲
Naiad damselfly

圖 / 李世仰（Photo：Shih-Yang Lee）

無霸勾蜓幼期水蠆及成蟲。
Naiad and adult jumbo dragonfly, *Anotogaster sieboldii*.

(三) 完全變態：看我七十二變

完全變態的昆蟲一生有卵、幼蟲、蛹、成蟲這四個階段，為人們所熟知的蝴蝶、蚊、蠅、蜜蜂、甲蟲等昆蟲皆屬於此類。

Holometamorphosis: Becoming totally different

Insects that undergo holometamorphosis have four life stages: egg, larva, pupa, and adult. This group of insect includes butterflies, mosquitoes, flies, bees, and beetles.

圖 / 蔡正隆（Photo：Cheng-Lung Tsai）

扁鍬形蟲生活史四階段：卵、幼蟲、蛹及雄成蟲。
Life cycle of the stag beetle *Dorcus kyanrauensis* includes egg, larva, pupa, and adult (male).

圖 / 吳怡欣（Photo：I-Hsin Wu）
圖 / 詹明澍（Photo：Ming-Shu Chan）

黃裳鳳蝶生活史四階段：卵、幼蟲、蛹及雌雄成蟲。
The life cycle of bird-wing butterfly, *Troides aeacus*, includes four stages: egg, larva, pupa, and adult, shown here are female and male (left).

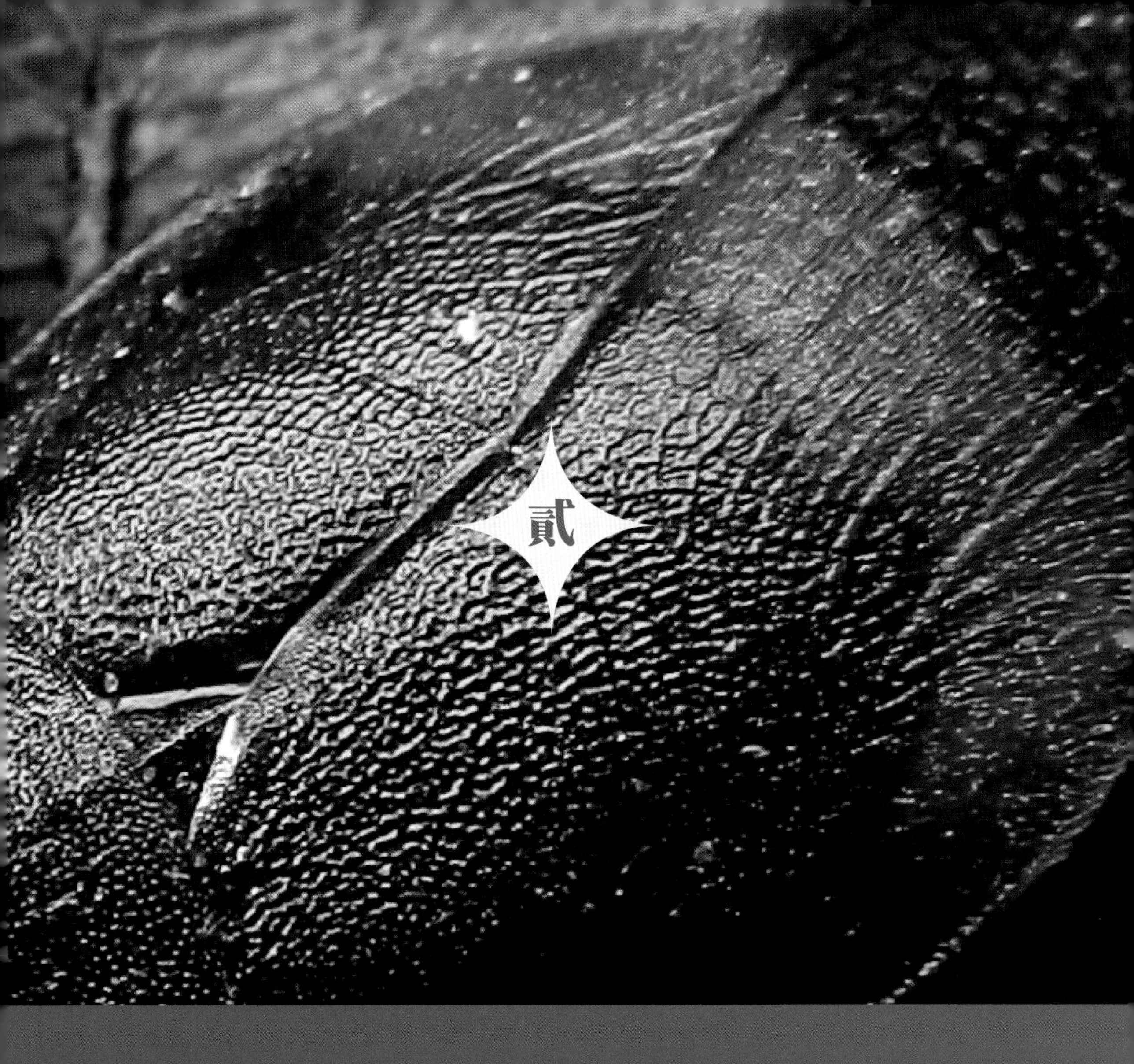

昆蟲的構造

Insect structure

葉文斌
Wen-Bin Yeh

二、昆蟲的構造

昆蟲外觀分節具外骨骼及副器構造，內部則具生理功能有關的各類組織系統，都與其存活繁衍有重要關係。

Insect structure

All insects have a segmented body, an exoskeleton, and jointed appendages. Inside the body, there are tissues and organs that work together to provide specific physical movements and physiological functions. These structures and functions enable insects to survive and reproduce.

(一)外部構造

昆蟲的外觀為頭、胸、腹三個部分組成，頭部有重要的副器，如口器、複眼及觸角；胸部 3 節各有一對足，中後胸多有一對翅；腹部 11 節，末節有尾毛。有些昆蟲體節具腺體，分泌特殊化學物質。

External structure

The insect body is divided into three parts: the head, thorax, and abdomen. The head contains vital organs such as the mouthpart, compound eyes, and antennae. The thorax has three pairs of legs, with one pair in each segment of the thorax: prothorax, mesothorax, metathorax. Two pairs of wings are usually found attached to the mesothorax and metathorax. The abdomen has 11 segments, the last of which has the cerci. In some insects, excretory glands secrete unique chemical compounds, which may be found on body segments.

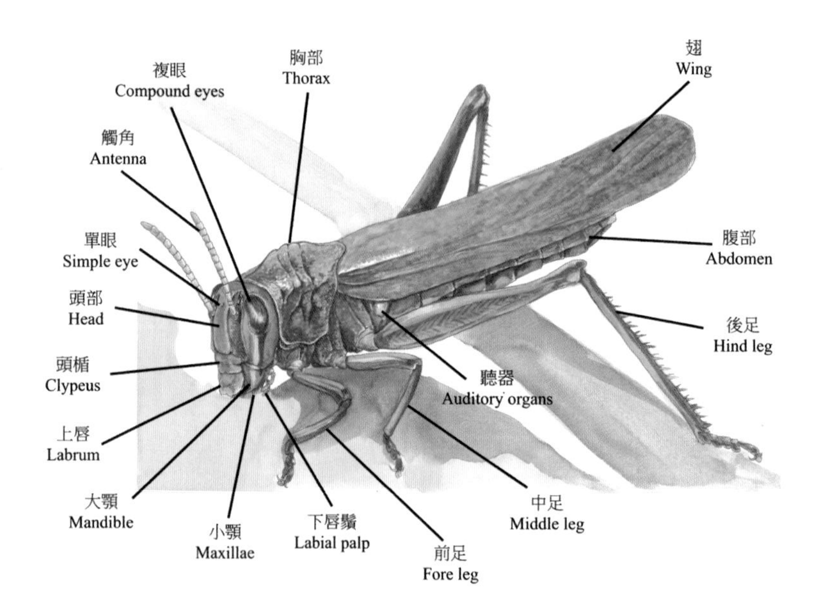

台灣大蝗示意的昆蟲外部重要副器及構造。
The external structure of an insect and its appendages, as shown in a Formosan giant locust.

(二) 內部構造

昆蟲內部主要構造除呼吸、消化、循環、排泄、神經、肌肉及生殖等系統及肌肉著生的內骨外，亦有特別的內分泌系統、唾液腺及脂肪體等構造。

Internal structure

In addition to the major internal systems, such as the respiratory, digestive, circulatory, excretory, nervous, muscular, reproductive and endoskeletal systems, there are the endocrine system, salivary glands, and fat storage system.

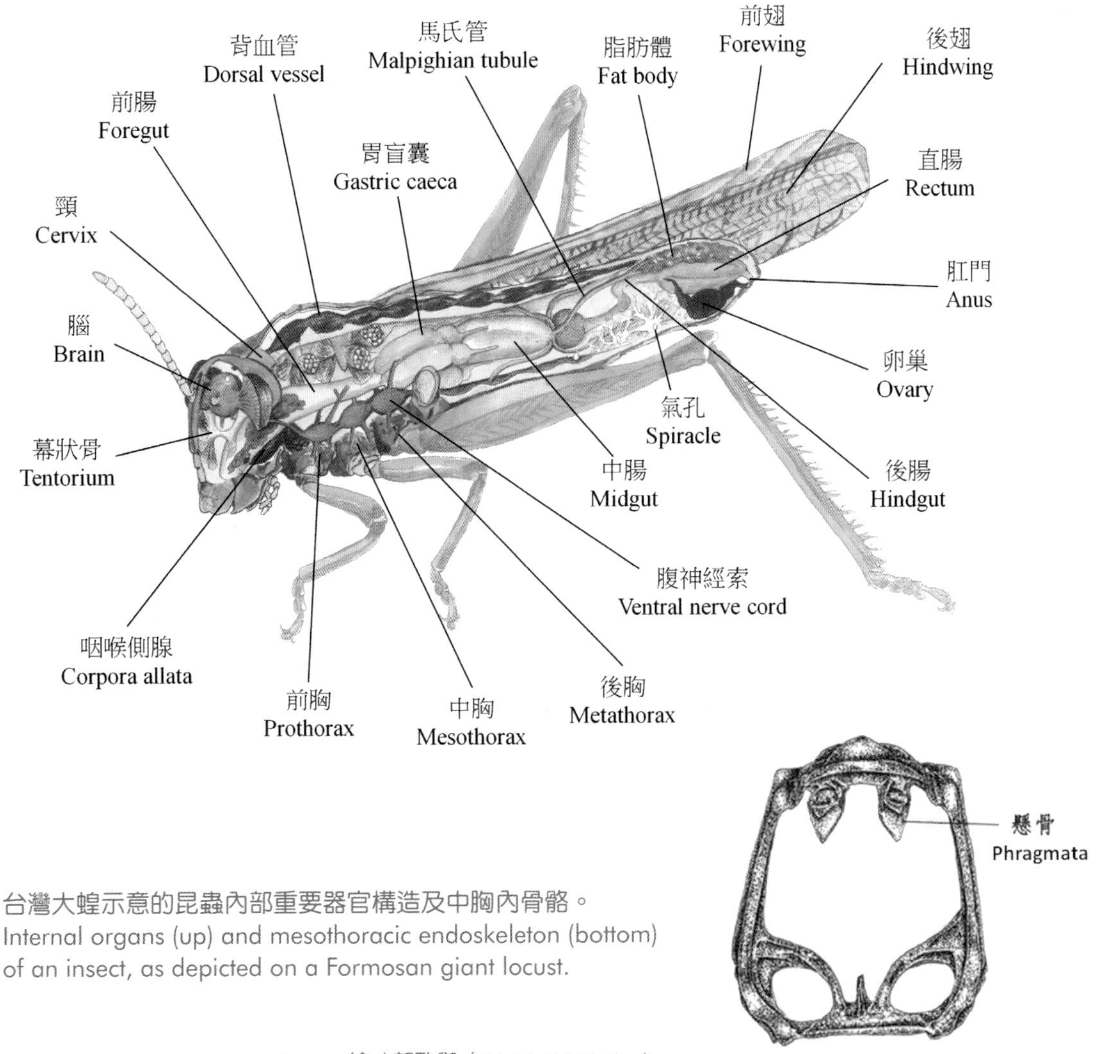

台灣大蝗示意的昆蟲內部重要器官構造及中胸內骨骼。
Internal organs (up) and mesothoracic endoskeleton (bottom) of an insect, as depicted on a Formosan giant locust.

1 | 2 繪 / 郭致與（Paint：Chih-Yu Kuo）

無所不在的昆蟲
上天、下地、潛水、入土

Insects everywhere
adaptive in air, ground, subterranean, and water

林鈺淳、趙宜閔、吳奕徵、葉文斌
Yu-Chun Lin、Yi-Min Chao、I-Chaen Wu、Wen-Bin Yeh

圖 / 廖一璋（Photo：Yi-Chang Liao）

水裡游的魚蛉幼蟲。
The dobsonfly larva dives in water.

三、 無所不在的昆蟲：上天、下地、潛水、入土

昆蟲的繁盛不是一天造成的，體型小加上多樣及特異的翅膀及附肢，得以適應各環境是大家較熟悉的要素；高效率的氣管系統、分節、外骨骼構造也是重要元素！不過，各種變態方法可能才是牠渡過種種環境難關的因素。

Insects everywhere: adaptive in air, ground, subterranean, and water

Evolution is the key driver of insect to become a dominant life-form on the earth. Insects have been highly adaptable to different environments because of their segmented small bodies, various forms of wing and appendage, highly efficient respiratory system, and exoskeleton. However, the ability of insects to undergo various types of metamorphoses may contribute the most in enabling insects to survive through environmental crises.

圖 / 楊曼妙（Photo：Man-Miao Yang）

空中飛行的食蚜蠅。
The hoverfly flies in the sky.

土裡鑽的山林原白蟻的工蟻及兵蟻。
The subterranean termite worker (left) and soldier (right) of *Hodotermopsis sjöstedti* nest underground.

地上爬的帶紋紺蠊。
The cockroach crawls on the ground.

1. 圖／黃宸瑋（Photo：Chen-Wei Huang） 1
2. 圖／周明勳（Photo：Ming-Hsun Chou） 2

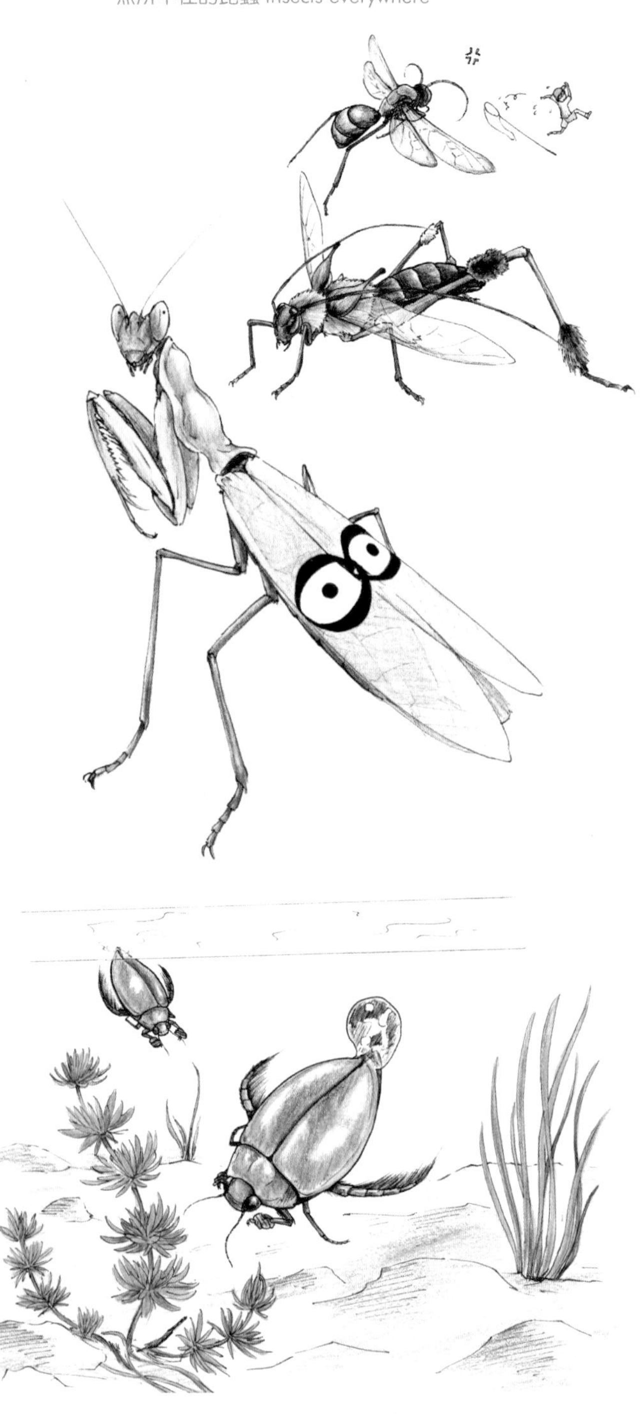

繪 / 郭致與（Paint：Chih-Yu Kuo）

上天、下地、潛水、入土，棲息於各類棲所的昆蟲。
By flying, walking, swimming, and hiding in the soil, insects have adapted to all kinds of habitats.

(一) 以昆蟲為師

四億多年的演化歲月下，昆蟲翱翔天際、悠游水中、潛藏大地；千變萬化的昆蟲，多樣特異，適應於各類環境，可說是無所不在，隨著這個地球運行。牠們的生存智慧值得我們省思，學習以昆蟲的角度，看見各式各樣的廣闊世界。

Learning from insects

Over the past 400 million years, insects have evolved to be exceptionally good at flying, swimming, and hiding in the ground. As a result of their evolution, insects are highly diverse and adapted to live almost everywhere in our world. Because insects play important roles in our planet, learning the way insects live allow us to better understand our world.

（二）陸地上的霸主：捨我其誰

「族繁不及備載」的昆蟲家族，歷經地貌、環境的巨大考驗，自有其生存之道。牠們高效率的氣管系統、具彈性的分節與附肢，與保護內臟、減少水份蒸散及提供肌肉附著的外骨骼等特殊構造之外，尚有適應各類環境的附器，如五花八門的觸角、不同功能的口器、花招百出的複眼、追趕跑跳碰的足、變化多端的翅，還有各式各樣產卵器；體型小及成長過程中的變態現象，更使牠們能在各類環境下適應生存。

Insects territorialize and dominate the planet

Insects have been said to be a group "too numerous to numerate" in the animal kingdom. They have found ways to survive in extreme and changing environments. For instance, insects possess highly sophisticated features, such as a highly efficient tracheal system, flexible segmented body, and exoskeleton. The exoskeleton not only protects the internal organs of an insect, but also reduces water evaporation, and enabling muscle attachment and body movement. In addition, insects have diverse appendages that adapt well to different environments. These appendages include the multi-functional antennae, mouthparts, compound eyes, legs, wings, and ovipositors. Moreover, insects' small size and ability to metamorphosize enable them to survive successfully in many different environments.

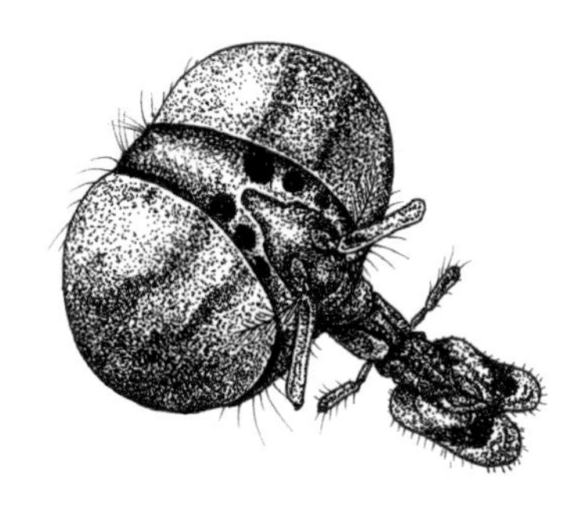

家蠅舐吮式口器
Lapping type mouthpart of the housefly

蝴蝶曲管式口器
Siphoning type mouthpart of the butterfly

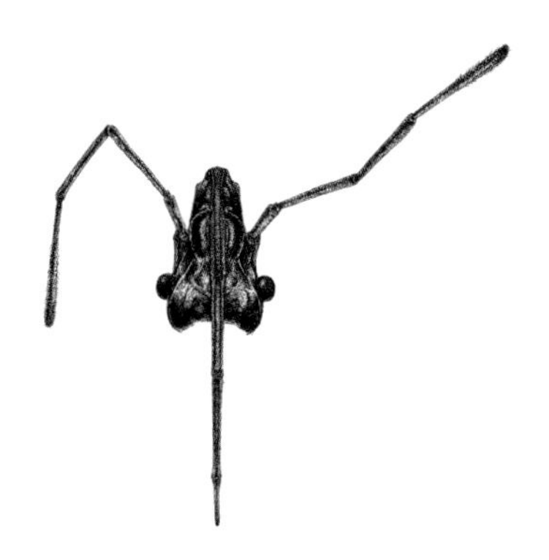

椿象刺吸式口器
Piercing-sucking type mouthpart of the stink bug

紅粉燈蛾羽狀觸角
Bipectinate-like antenna of the tiger moth

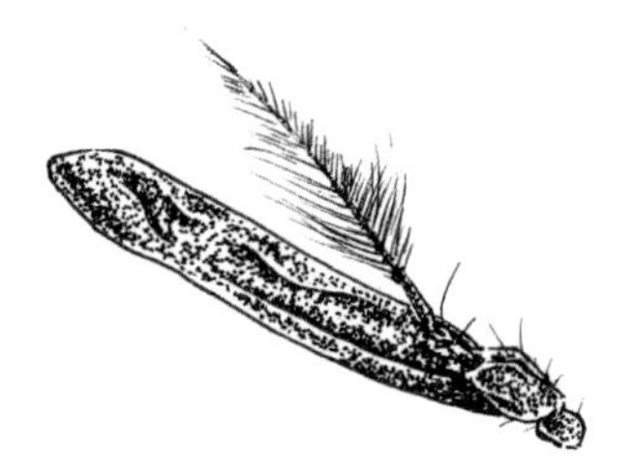

家蠅不正形觸角
Irregular antenna of the housefly

青銅金龜腮葉狀觸角
Lamellate-like antenna of the scarab beetle

蝗蟲跳躍足
Jumping leg of the locust

螳螂捕捉足
Capturing leg of the mantis

蜜蜂攜粉足
Pollen-carrying leg of the bee

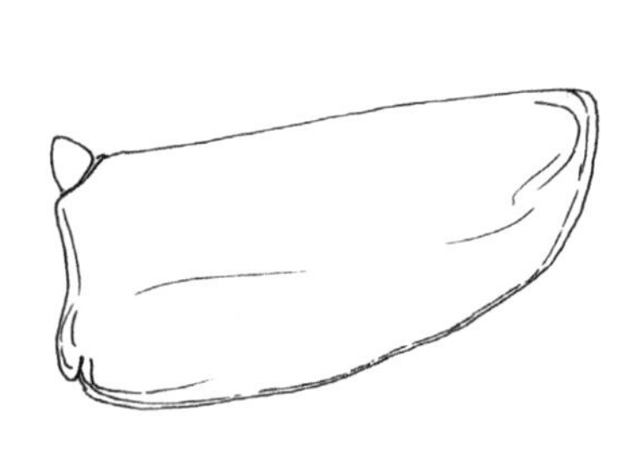

甲蟲翅鞘
Elytra of the beetle

蛾類的翅
Scale wing of the moth

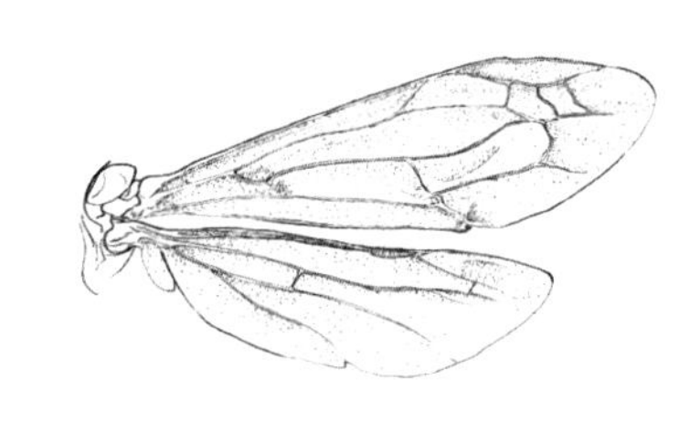

胡蜂膜翅
Membranous wing of the hornet

繪 / 郭致與（Sketch：Chih-Yu Kuo）

昆蟲的口器、觸角、足、翅等可適應各類環境棲所的副器。
Insects have diversified accessories of mouthpart, antenna, leg, and wing that help them to adapt to different habitats.

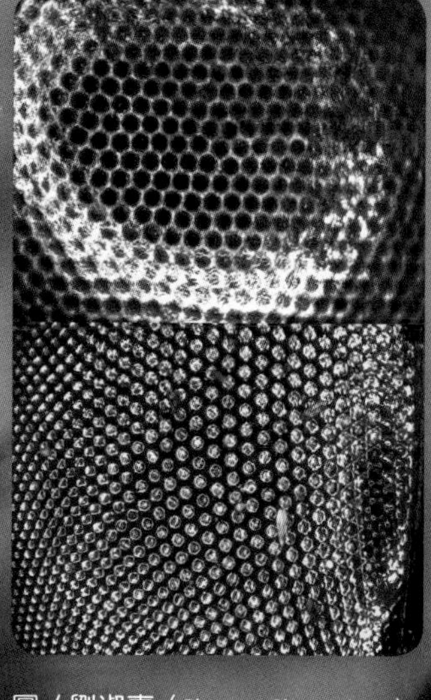

圖 / 劉淑惠（Photo：Shu-Hui Liu）

食蟲虻複眼及其細微小眼。
Aslid compound eyes and its ommatidia.

(三) 空中獵捕高手

蜻蜓、食蟲虻及虎頭蜂是空中獵捕的昆蟲代表，藉助優良的飛行能力與碩大的複眼，成功聚焦追獵；此空中生態中，搖蚊、斑蚊、葉蟬及蟬為常被捕食代表。

圖 / 江允中（Photo：Yun-Chung Chiang）

姬蜂虻
Bee flies, *Systropus*

A flying hunter

Dragonflies, robber flies, and hornets are flying insects that hunting prey in the air. With excellent flying skills and well-developed compound eyes, they focus and lock the prey on from far. They usually prey on non-biting midges, mosquitoes, leafhoppers, and cicadas.

數萬個小眼組成的發達複眼，讓不少昆蟲得以進行空中獵捕。

The developed compound eyes which are composed of tens of thousands of ommatidia that allow many insects to hunt in the air.

圖 / 黃致玠（Photo：Chih-Chieh Huang）

霜白蜻蜓 (上雄及下雌)
Dragonfly, *Orthetrum pruinosum*, Male (above), Female

(四)生機盎然的土表樂園

森林底層的枯落葉雖非綠意盎然，但可是個生氣勃勃的世界；眾多跳蟲、嚙蟲、步行蟲、蜚蠊、等棲息於此，也是清道夫糞金龜、埋葬蟲的世界，更是眾多昆蟲隱匿休息的地方。

Life in top soil: A paradise

Although the leaf litters on the forest floor are not green, they are home to springtails, bark louses, ground beetles, and cockroaches etc. Also, they are home to scavenging dung beetles and carrion beetles, beside it is a hide-out or rest-area for many insects.

圖 / 蔡正隆（Photo：Cheng-Lung Tsai）

地表是直翅類螽蟴、蟋蟀及蝗蟲重要棲所。
The ground surface is a vital habitat for orthopterid katydids, crickets, and locusts.

繪 / 朱能榮（Paint：Neng-Jung Chu）

森林底層是眾多昆蟲的遊樂園。
Forest floor is the paradise for many insects.

（五）「混」在水裡「混」

昆蟲的一生不論在幼生期或是成蟲期的任一階段，生活在水裡，都可稱〝水生昆蟲〞。既然在水裡混便有著異於常〝蟲〞的呼吸方式及構造，有攜著氣泡的龍蝨，有著氣管鰓的蜻蛉、蜉蝣及石蠅的稚蟲，有著呼吸管的紅娘華等等。湍急、速緩的溪流、平靜的水窪、水塘、樹洞等，只要是有水的環境幾乎都是牠們的家，連髒水也可存活；因此，水生昆蟲亦可用來做為監測水質的受汙染程度的指標生物喔！

Aquatic insects: the best diver

Aquatic insects complete their life cycle (or partly) in the water. To live in water, they have special structure for respiration. For example, adult diving beetle uses air-bubble, naiads of damselflies, mayflies, and stoneflies use tracheal gills, while water scorpions use respiration siphons, to exchange air. They can be found at water bodies, water ways, a rushing or slow running streams, puddle, pools or tree holes. They also can be found in the turbid and dirty water body. Therefore, aquatic insects are well known as bioindicators to monitor the water quality and degree of water pollution.

繪／朱能榮（Paint：Neng-Jung Chu）

城市污水、湖泊及溪流住有各類水棲昆蟲。
Dirty ditches, calm lakes, and running streams are inhabited by different kinds of aquatic insects.

(六) 潛藏於地下的王者

對人類來說暗無天日的土表層，可是白蟻、螞蟻悠游的天地，也是獨角仙隱居的地點；在枯倒木下，更是眾多土棲昆蟲巡視、狩獵、繁衍的樂園。

Subterranean insects

The topsoil is a habitat that dominated by termites and ants. It is a hiding place for rhinoceros beetles as well. The insects search food and rear their babies where the deadwood and the topsoil are closely connected.

繪 / 朱能榮（Paint：Neng-Jung Chu）

土表層下是各類土棲昆蟲的潛藏樂園。
There is a hiding world under topsoil for subterranean insects.

昆蟲的生殖適應
Reproductive strategy of insects

雷喬安、葉文斌
Chaou-An Rei、Wen-Bin Yeh

繪 / 劉淑惠（Paint：Shu-Hui Liu）

亮麗迷人的黃裳鳳蝶。
The dazzling and charming golden birdwing butterfly, *Troides aeacus formosanus*.

四、 昆蟲的生殖適應

成功繁衍出下一代是生物能否在地球上存續的關鍵，演化出各式各樣的生殖策略，是昆蟲種類繁複多樣的重要原因。

Reproductive strategy of insects

Reproducing offspring successfully is essential to keep the species from extinction. Evolutionary change in reproductive strategies is the key making insect a highly diverse group of living organism.

振翅呼喚的扁頭蟋蟀。
The flat-headed cricket, *Loxoblemmus appendicularis*, calls by wing vibration.

放閃螢光的山窗螢。
The firefly, *Pyrocoelia praetexta*, flashes to attract the female.

1/2 圖 / 詹明澍（Photo：Ming-Shu Chan）

（一）求偶：「我們結婚吧！」

在人類的世界中，當男生與女生相愛，即有可能結婚，這之間有名為「愛情」的奇妙元素。但在昆蟲的世界裡，光是依靠萍水相逢、巧遇，可能是不夠的；昆蟲該如何展現魅力、吸引異性呢？

Insect courtship: "Would you marry me?"

Marriage may happen when man and woman fall in love. Unlike human, simply encountering may not be enough for a successful courtship. How does an insect show its charm and attract its ideal mate?

圖／鄒議慷（Photo：Yi-Kang Tzou）

有備而來的灶馬。
The spotted camel crickets, *Ceuthophilus maculates*, offer food to his mate.

1. 一見鍾情 — 視覺求偶

蝴蝶翅膀上的美麗花紋、螢火蟲的閃爍螢光都可吸引異性前來。

Love at first sight – Visual display

The beautiful patterns on butterfly wings and the bioluminescence of firefly are used to attract and to find their mates.

圖 / 廖智安（Photo：Chi-An Liao）

孔雀青蛺蝶藉由其獨特的亮麗外形吸引異性。
The peacock blue butterfly, *Junonia orithya*, attracts mate with its unique and colorful appearance.

圖／黃致玠（Photo：Chih-Chieh Huang）

黃胸黑翅螢藉由發光吸引異性，前來配對。
The blinking signals that the firefly, *Aquatica hydrophila*, send out are intended to attract mates.

圖 / 江允中（Photo：Yun-Chung Chiang）

雄草蟬藉由腹基前方的鼓膜發音器振動發音，吸引異性。
The male grass cicada, *Mogannia hebes*, produces sound through a pair of special sound-producing organs or "tymbals" located on the basal ventral abdominal segment.

2. 一聽鍾情 — 聽覺求偶

蟋蟀、蝗蟲及蟬等鳴蟲，藉由發聲鳴叫招引異性。

Finding love through melodies — Auditory courtship signal

Acoustic communication in insects is used as courtship signal. Insects including crickets, locusts, and cicadas produce the signal to attract the mates.

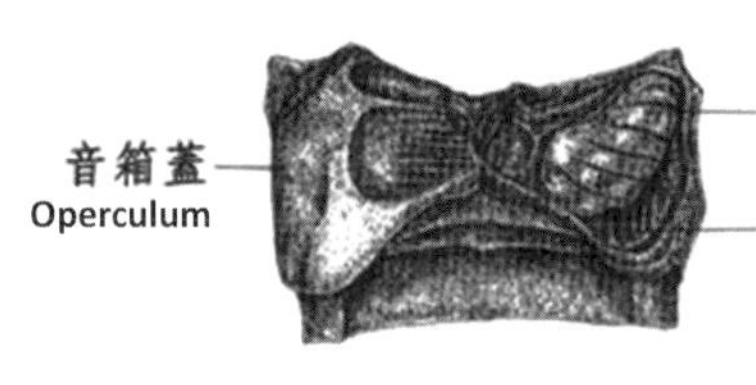

繪 / 郭致與（Sketch：Chih-Yu Kuo）

蟬聽器
Tympanum organ of cicada

圖 / 蔡正隆（Photo：Cheng-Lung Tsai）
圖 / 詹明澍（Photo：Ming-Shu Chan）

扁頭蟋蟀藉由左右翅上的發音器摩擦發聲。
The flat-headed cricket, *Loxoblemmus appendicularis*, produces sound by rubbing both left and right wings.

繪 / 朱能榮（Paint：Neng-Jung Chu）
圖 / 雷喬安（Photo：Chaou-An Rei）

台灣大蝗藉由翅腿的摩擦發聲，並藉由腹基背側聽器接收聲音。
The Formosan big locust, *Chondracris rosea*, creates sounds by stridulation which is the rubbing of wings and legs; and received by tympanal organ locating in the dorsal lateral of the abdominal base.

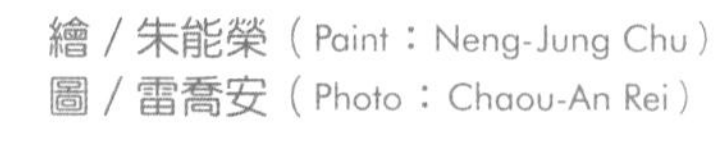

3. 結婚送聘禮 — 餽贈求偶

雄性蚊蠍蛉將獵物送給雌蠍蛉作為聘禮，以求芳心。

Nuptial gifts — A bribe for bride

The male scorpionfly offers a prey as a nuptial gift to lure his mate for copulation.

繪 / 陳姵雯（Paint：Pei Win Chen）

雄蚊蠍蛉準備夠分量的食物送給雌蠍蛉，以求芳心。
The male scorpion flies offer foods to the female for mating.

圖 / 廖智安（Photo：Chi-An Liao）

鬼豔鍬形蟲遇到求偶的競爭對手，常需奮戰拼鬥一番。
The stag beetle, *Odontolabis siva*, fights hard when it meets a competitor.

4. 為愛而戰 — 競爭求偶

有些甲蟲常靠戰鬥贏取新娘，也捍衛自己的愛。

Fight for mate — A competition courtship

Some species of beetles fight for their right to mate.

圖 / 黃致玠（Photo：Chih-Chieh Huang）

大型獨角仙有體型優勢，較可能在眾多雄性中脫穎而出。
With a size advantage, large rhinoceros beetle, *Trypoxylus dichotomus*, is more likely to outcompete the other males.

（二）交配：「你是我的真愛」

當蟲爸與蟲媽順利求偶成功後，就要進入「交配」階段。小小的昆蟲怎麼確定找到的就是真愛呢？具有外骨骼的昆蟲有一「Key & Lock」假說 - 正確的鑰匙才能打開相對應的鎖，也從而發展出多樣的交配方式。

Copulation: "You are my true love"

Going through a series of courting behaviour, two insect lovers go to "copulation". How does insect know he/she has found the right one? There is a hypothesis so called "Key & Lock" that only the right key can unlock the lock. Combinations of key and lock in insect copulation performances were therefore been developed.

黃腹鹿子蛾多呈線型尾對尾的交尾狀況。
The tail-to-tail copulation posture is frequently found in *Amata perixanthia*.

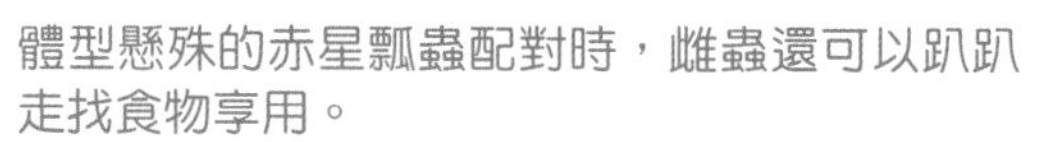

體型懸殊的赤星瓢蟲配對時，雌蟲還可以趴趴走找食物享用。
During copulation, the large-size female ladybug, *Lemnia saucia*, carries the male on her back and searching for food.

豆芫菁配對時，觸角會緊緊纏繞。
When the beetles *Epicauta hirticornis* mate, their antennae are intertwined.

腹部末端具有特化把握器的青紋絲蟌，會攫住自己的愛。
Male damselfly, *Indolestes cyaneus*, has an abdominal specialized clasper to hold the female tightly.

1. 圖 / 江允中（Photo：Yun-Chung Chiang）
2. 圖 / 周明勳（Photo：Ming-Hsun Chou）
3. 圖 / 江允中（Photo：Yun-Chung Chiang）
4. 圖 / 詹明澍（Photo：Ming-Shu Chan）

	1	
	2	3
	4	

1. 精力大補丸

螽蟴、蟋蟀利用精胞傳遞精子，也作為雌蟲的食物。

A nutritious spermatophore

In male katydids or crickets, spermatophore does not only carry sperms, but also provide food to his mate.

圖 / 黃致玠（Photo：Chih-Chieh Huang）
圖 / 詹明澍（Photo：Ming-Shu Chan）

直翅目的螽蟴及蟋蟀經由外掛精包於雌性生殖器的方式進行受精，精包富含營養物質可供雌蟲享用。
Orthopteroid katydids and crickets undergo fertilization by transferring elaborate spermatophore which is attached externally to the female's genitalia. The nutrient-rich spermatophore is eaten by the female after mating.

2. 危險的婚禮

雄螳螂在交配時是命懸一線的，可能為愛獻出自己寶貴的生命。

A dangerous wedding

Male mantis risks his life in copulation, he may sacrifice himself as food for the female.

圖／黃致玠（Photo：Chih-Chieh Huang）

螳螂演化出特別的交配行為，雄蟲會奉獻出自己的頭及身體給雌蟲作為食物。
Mantis evolves a peculiar mating behavior in which the male would provide its head and body to female as food while mating.

3. 愛的圓舞曲

　　蜻蜓及豆娘交配時會勾勒出象徵「愛」的美麗心形。

A heart-shaped mating posture

　　Dragonfly and damselfly form the heart-shaped posture when mating.

圖 / 黃致玠（Photo：Chih-Chieh Huang）
圖 / 江允中（Photo：Yun-Chung Chiang）

蜻蛉目的豆娘及蜻蜓交配時，會呈現愛心型的美麗姿勢。
When damselflies and dragonflies mate, a beautiful heart-shaped posture is formed.

4. 愛情疊疊樂

金花蟲交配時常常是不甘寂寞的。

Pile up for love

Male leaf beetles usually appear in group when copulating .

圖 / 周明勳（Photo：Ming-Hsun Chou）

緬甸藍葉蚤雄蟲競相爭取第一的配對機會，會有疊在一起的特殊姿勢。
Males of blue leaf beetles, *Altica birmanensis*, usually pile up to compete the first mating candidate.

圖 / 李世仰（Photo：Shih-Yang Lee）

鐵甲蟲身為金花蟲的一員，雖全身是刺，配對時仍有疊疊競爭的本色。
The leaf beetles, *Dactylispa sauteri*, though the body is thorny, they pile up to compete for a chance of copulation.

（三）產卵：「生生不息」

多樣特異的昆蟲世界中，演化出各樣的產卵方式。多數昆蟲都像蝴蝶般卵生，將卵產在幼體可存活的棲所；有些昆蟲則是直接產出幼體的胎生，牠們演化出不同的繁殖方式，讓種族延續。

Eggs reproduction: "Circle of life"

Diverse and extraordinary insects have evolved various ways to lay eggs. Most of them are ovipara, like butterflies laying eggs at habitats where resource is plenty. While the viviparous insects give birth to young ones. They have evolved differently in reproductive strategies for the continuity of the species.

圖／詹明澍（Photo：Ming-Shu Chan）

雲杉球蚜產卵後，若蟲會住入其刺激寄主所產生的癭室。
The spruce ball aphid, *Adelges* sp., lays its egg on spruce and the hatched nymphs will inject stimulant to the host to produce gall which is the incubator for the hatched nymphs to live inside.

圖／廖啟淳（Photo：Levic Liao）

寄生蜂產卵於青枯葉蛾，其幼蟲吃穿不愁。
The parasite wasp lays its eggs on the moth, *Trabala vishnou*. Therefore, the larvae have plentiful supply of food and a safe shelter.

圖 / 江允中（Photo：Yun-Chung Chiang）
圖 / 鄒議慷（Photo：Yi-Kang Tzou）

蚜蟲的卵體內孵化後，直接生出成長的幼體。
Female aphid produces young aphid as the egg is hatched inside the female body.

1. 超級複製蟲 — 多胚生殖

具有全功能特性，從卵細胞分裂來的卵都可發育為一個個體；在一些小型的繭蜂、細蜂、釉小蜂、螯蜂及撚翅蟲都可見。

Super cloning — Polyembryony

Two or more embryos develop from a single egg and give rise to numerous offspring. Common species of holo-embryony insects are small braconids, platygastrids, encyrtids, dryinids. and twisted-wing parasites.

繪 / 劉淑惠（Paint：Shu-Hui Liu）

少數寄生性的昆蟲胚胎如幹細胞一樣成長，分裂的卵都可發育成一個個體。
Embryos of some parasitic insects are divided like what the stem cells do, every divided embryo potentially develops into an individual.

2. 早生貴子 — 幼體生殖

幼蟲時期的卵細胞即已成熟，可產出子代；如美國隨木材貿易擴散到世界各地的微弱筒蠹蟲 (*Micromalthus debilis*)。

Young mommy — Paedogenesis

Paedogenesis is the reproduction by larvae or juveniles. In insects this form of reproduction is known in American *Micromalthus debilis* which makes its descendants spread over the world by wood trading.

繪 / 郭致興（Paint：Chih-Yu Kuo）

奇特的微弱筒蠹蟲，有時在幼體的外型時，即具有生殖能力。
Sometimes, *Micromalthus debilis* beetle reproduces in an immature stage

3. 體內孵蛋 — 卵胎生

卵於母體的生殖道內發育，至孵化時才從母體產出。

Internal incubation — Ovoviviparity

Ovoviviparity, in which fertilized eggs containing yolk and enclosed in some form of eggshell are incubated in the reproductive tract of the female.

圖 / 周明勳（Photo：Ming-Hsun Chou）
圖 / 唐立正（Photo：Li-Cheng Tang）

衛生害蟲肉蠅是卵胎生，直接產出的蛆常是餐廳被客訴的原因。
Hygiene pest such as maggot fly, *Sarcophaga* sp., is oviviparous, it can produce maggot directly by which is often complained by restaurant customers.

圖 / 廖智安（Photo：Chi-An Liao）

邊吃邊生的蚜蟲直接生出一堆小蚜蟲，是牠成為農民頭痛害蟲的重要原因。
Sometimes, aphids give a lot of birth to their young while feeding on crops. This causes serious pest problems for farmers.

4. 單親媽媽 — 孤雌生殖

Single mother — Parthenogenesis

未受精的卵可發育成新個體。

Individual developed from unfertilized eggs is called parthenogenesis.

圖 / 游依靜（Photo：Yi-Ching Yu）

孤雌生殖是很多昆蟲傳宗接代的秘訣，津田氏大頭竹節不需配對即可產生大量的卵。
Parthenogenesis occurs commonly among insects. Tsudai big-head stick insect can produce a large number of eggs without mating.

大田鱉爸爸守在雌蟲所產的卵塊旁，隨時呵護照顧。
Male giant water bug guards the eggs laid by the female and takes care of them all the time.

(四)護幼：「愛寶貝」

在人類的世界裡，父母照顧子女是理所當然的事，但在昆蟲卻不一定是如此。多數的昆蟲將卵產在適當的棲所後就離去了，但有些昆蟲則如人類般演化出護卵及育幼的特殊行為。

Parental care: "Nursing eggs and larvae"

In human, parental care is congenital yet this may not be necessary for insects. Most insects leave their eggs alone after placing them at a safe place; however, some insects evolve parental care to their offspring such as protecting eggs and nursing babies.

推糞金龜產卵於自製糞球內，不僅將糞球推到地下窩內保護起來，還讓孵化的幼蟲有充足食物。
The dung beetles lay eggs in the dung ball and push the dung ball in the underground for protection. In the meantime, the dung ball can be an immediate supply of food to the young once hatched.

如同其牠螞蟻，琉璃蟻將後代保護於巢內。
Like other ants, *Dolichoderus* ants keep their offspring in their nests for protection.

昆蟲界的模範父親負子蟲背著卵，可隨時保護雌蟲所產的卵。
The giant water bug is one of the typical examples in insect that the male is tasked with caring for the eggs. He protects the eggs all the time.

1. 圖 / 廖智安（Photo：Chi-An Liao）
2. 圖 / 江允中（Photo：Yun-Chung Chiang）
3. 圖 / 江允中（Photo：Yun-Chung Chiang）
4. 圖 / 鄒議慷（Photo：Yi-Kang Tzou）

1 2 3 4

1. 愛寶貝

蠼螋細選巢室保護幼蟲至可獨立生活、椿象將卵產在植物上細心呵護、負子蟲背卵隨時保護。

Caring babies

Earwigs carefully select nests and protect their larvae until they grow up and be independent. Stinkbugs lay their eggs on plants and take good care of them. Water bugs carry eggs on the back for protection until the eggs hatched.

有不少種類的昆蟲會細心照護後代，確保自己的寶寶可以存活下來。

Many insects can take good care of their offspring to ensure that their babies can survive throughout.

圖 / 翁逸明（Photo：Yi-Ming Weng）

蠼螋媽媽護卵及護幼。
Eggs and young are protected by earworm mother.

圖 / 唐昌迪（Photo：Chang-Ti Tang）

負子蟲爸爸護卵。
Eggs are protected by male water bug.

圖 / 黃致玠（Photo：Chih-Chieh Huang）

椿象媽媽護卵及護幼。
Eggs and young protected by stinkbug.

2. 愛的防護罩

螳螂及蟑螂將卵產在螵蛸中保護。

A protective shield

Mantis and cockroaches lay their eggs in the ootheca which provides a protective outer covering to protect the eggs.

蟑螂雖令人討厭，與牠近親的螳螂都有演化出其獨特護卵的螵蛸。

Although cockroach is an annoying pest, while it and its close relatives mantis have evolved a unique reproductive method, ootheca, to protect their eggs.

圖 / 廖一璋（Photo：Yi-Chang Liao）

螳螂及其螵蛸。
Mantis and its ootheca.

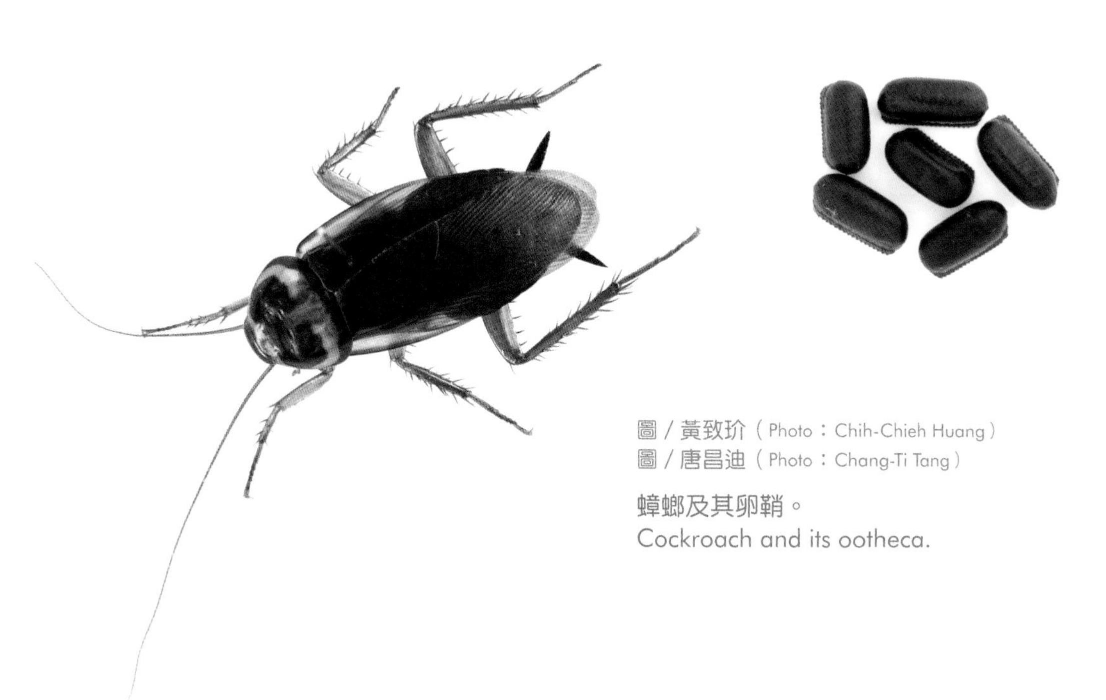

圖 / 黃致玠（Photo：Chih-Chieh Huang）
圖 / 唐昌迪（Photo：Chang-Ti Tang）

蟑螂及其卵鞘。
Cockroach and its ootheca.

3. 活動育嬰室

將卵產在獵物體內，讓幼蟲不愁溫飽。寄生蜂產卵在蝶蛾幼蟲體內，小寶貝在裡面取食，隨著這幼蟲四處走動直到羽化。

A moving nursery room

Lay eggs onto their prey, so that the emerged larvae have ample food to feed on before emergence. The parasitic wasps lay eggs in larvae of the moth and butterfly. The emerged larvae consume the host tissues, transported by the host until they emerge to adult.

圖 / 廖一璋（Photo：Yi-Chang Liao）

被寄生蜂寄生的蛾類幼蟲。
The caterpillar of moth is parasitized by the parasitic wasp.

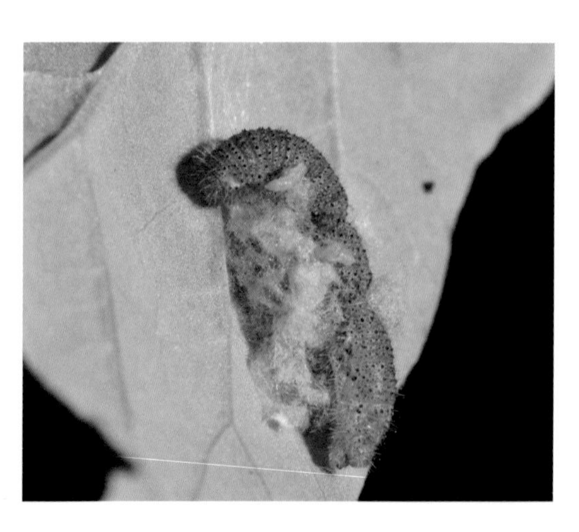

圖 / 鄒議慷（Photo：Yi-Kang Tzou）

被寄生蜂寄生的蝶類幼蟲。
The caterpillar of butterfly is parasitized by the parasitic wasp.

4. 飽暖育嬰室

有些昆蟲會產卵於牠精心製作的育嬰室，讓孵化的幼蟲吃住不愁。搖籃蟲媽媽產卵於精心製作的巢室、糞金龜產卵在特製糞球中，都讓小孩有可溫飽的家。

A perfect nursery

Some insects lay their eggs in a well-constructed nursery, allowing the young to have enough food and comfortable place to stay. Female leaf-rolling weevils protect newly laid eggs by rolling them up inside a leaf. Dung beetles lay eggs in their dung-ball. Both of them give their children a comfortable nursery.

圖 / 李世仰（Photo：Shih-Yang Lee）

糞金龜及其育嬰糞球。
Dung beetle holds the dung ball which is nursery home for her larvae.

圖 / 黃致玠（Photo：Chih-Chieh Huang）
圖 / 劉淑惠（Photo: Shu-Hui Liu）

棕長頸捲葉象鼻蟲會折葉製做育嬰搖籃。
The brown giraffe weevil, *Paratrachelophorus nodicornis*, would fold leaf to accommodate offspring.

圖 / 江允中（Photo：Yun-Chung Chiang）

黑點捲葉象鼻蟲及其育嬰搖籃。
The black spotted giraffe weevil, *Agomadaranus pardaloides*, makes a nursery cradle.

伍

昆蟲的變裝秀

Insect dressing in evolutionary game

吳奕徵、葉文斌
I-Chaen Wu、Wen-Bin Yeh

五、 昆蟲的變裝秀

昆蟲的隱蔽、偽裝、擬態多樣特化，不管是混於環境讓你找不到、融入背景裝可愛、擬態成不可侵犯或直接告訴你別惹我，在在讓生命的演化大戲處處精彩動人。

Insect dressing in evolutionary game

Insects have evolved adaptive tactics to cheat their identity. These adaptations include camouflage, disguise, and mimicry. They blend in perfectly with the colors and shapes of its environment, and mimic predators to scare them away. The evolutionary consequences of the insect have been fascinating us.

圖 / 詹明澍（Photo：Ming-Shu Chan）

圓翅紫斑蝶
Blue-banded king crow, *Euploea eunice*

(一) 穆氏擬態：猜猜我是誰！

物種間外型相仿，大力宣告我們有毒不要吃；有一點、藏三點、亂亂點等白斑點點的端紫、圓翅、斯氏、小紫斑蝶即為一例。

Müllerian mimicry: Guess who I am!

Some insects are morphologically similar each other which advertizing them are unpalatable. Purple crow butterflies in Taiwan, e.g. *Euploea mulciber, E. eunice, E. sylvester*, and *E. tulliolus*, are the group, for example, retaining similar white spots and color pattern on wings to demonstrate that they are unpalatable.

圖 / 葉文斌（Photo：Wen-Bin Yeh）

斯氏紫斑蝶
The double-banded crow, *Euploea sylvester swinhoei*

圖 / 李世仰（Photo：Shih-Yang Lee）

小紫斑蝶
Small purple crow butterfly, *Euploea tulliolus koxinga*

圖 / 黃致玠（Photo：Chih-Chieh Huang）

端紫斑蝶
The striped blue crow, *Euploea mulciber barsine*

台灣數種紫斑蝶外形彼此相似，雖傳承祖先，也與穆氏擬態有關。

The appearances of purple crow butterflies in Taiwan are inherited from their common ancestors. They may be similar because of Müllerian mimicry.

圖 / 李世仰（Photo：Shih-Yang Lee）

外觀像蜂的東方果實蠅。
The oriental fruit fly, *Bactrocera dorsalis*, evolves the appearance that resembles the hornet.

（二）貝氏擬態：披著羊皮的狼在昆蟲

不具危險性的物種模仿具危險性的物種，如無毒的雌紅紫蛺蝶跟有毒的樺斑蝶就十分相似；超級大害蟲東方果實蠅模仿兇猛的虎頭蜂也是演化絕品。

Batesian mimicry: A goat with wolf's skin

There are some non-poisonous insects mimic the poisonous species. For example, non-poisonous *Hypolimnas misippus* butterfly looks similar to the poisonous *Danaus chrysippus*. The oriental fruit fly is another excellent mimicker of hornets.

圖 / 江允中（Photo：Yun-Chung Chiang）

外觀像蜂的虻。
The fly evolves the appearance resembling the hornet.

圖 / 徐玉玲（Photo：Yu-Ling Hsu）

外觀像蜂的南瓜實蠅。
The pumpkin fly, *Bactrocera tau*, evolves the appearance that looks like the hornet.

圖 / 詹明澍（Photo：Ming-Shu Chan）

樺斑蝶幼蟲取食摩羅科植物，會將強心配醣體類毒素儲存於脂肪體。
The fat body of adult *Danaus chrysippus* stores the toxic compound of calactin, which is acquired by its larva when eating milkweed plants.

圖 / 詹明澍（Photo：Ming-Shu Chan）

體未含有毒性物質，但模擬有毒性樺斑蝶外觀的雌紅紫蛺蝶。
The nontoxic butterfly, *Hypolimnas misippus,* mimics the poisonous milkweed butterfly, *Danaus chrysippus.*

具毒武器的虎頭蜂及深藏劇毒的斑蝶可保護自己，是自然界常見物種；其他昆蟲會模仿牠們外形以保護自己。

Hornets and Danaid milkweed butterfly possess the poisonous weapons and toxic body contents as defense mechanisms. Other insects mimic the appearances of these insects to create a false alarm to other predators.

（三）隱蔽高手：來啊！你看不到？

不少昆蟲演化出保命的形態，可隱蔽於環境讓你找不到或融入背景裝可愛；竹節蟲、螽蟴及葉蟬即是此方面的隱藏大師。

Master of camouflage: Hide and seek?

Many insects may hide themselves in the environment to fool your eyes, blend themselves perfectly in with the background. These camouflage masters include stick insects, katydids, and leaf hoppers.

圖 / 詹明澍（Photo：Ming-Shu Chan）

蓬萊棘螽蟴具有隱蔽於棲息環境的綠色及褐色鑲嵌外觀。
Katydid, *Trachyzulpha formosana*, hides itself in its habitat with color of green and brown to blend in with their backgrounds.

演化出與環境相似的外形以避免被捕食，常見於各類昆蟲中。

Evolve an appearance that resembles the environment to avoid predation is common in the insect's world.

圖／黃宸瑋（Photo：Chen-Wei Huang）

綠色的葉蟬若蟲藏匿於綠色葉子的環境背景下。
Leafhopper's nymph with green color uses camouflage to match their green environment.

（四）偽裝高手：狼來了！介殼蟲沒！

部分昆蟲演化出它種昆蟲外形，可混入其中，獲得保護並有充分食物來源。

Deceive your foe: I'm good at disguising

Some insects have evolved the appearance resembling other insects which deceive those insect in order to allow themselves to live in the group for protection and food supply.

蚤蠅模擬白白嫩嫩的小白蟻，寄居窩內免費用餐；孟氏隱唇瓢蟲幼蟲特化出粉介殼蟲般的外型構造，混入蟲群中，盡情享用美食。

Phorid fly mimics young termite in order to live and feed at the termite nest. Ladybird's larvae of *Cryptolaemus montrouzieri* masquerades as mealybugs in appearance to foist themselves into the mimickees, thereby enjoying a big feast.

圖 / 梁維仁（Photo：Wei-Ren Liang）

蚤蠅特化外形易於藏入台灣土白蟻窩中。
The specialized appearance of phorid fly disguises its identity in the nest of fungus-growing termites, *Odontotermes formosanus*.

1 | 2
1. 圖 / 唐立正（Photo：Li-Cheng Tang）
2. 圖 / 葉文斌（Photo：Wen-Bin Yeh）

埃及吹棉介殼蟲及像超級大介殼蟲的群聚體。
Mealybug, *Icerya aegyptiaca*, and the big cluster of grouping mealybugs.

圖 / 莊益源（Photo：Yi-Yuan Chuang）

孟氏隱唇瓢蟲成蟲（左圖），其幼蟲模擬介殼蟲外形捕食介殼蟲，右兩圖分別為幼蟲背面及腹面。
Adult of ladybird, *Cryptolaemus montrouzieri* (left). Its larva is known as a mealybug predator which mimics mealybug. Dorsal (up) and ventral view (bottom) of ladybird's larva.

（五）警戒色：警告你！別惹我！

演化出瓢蟲警戒色的金花蟲。

Warning coloration: Don't touch me!

Some leaf beetles have evolved with ladybird's appearance so that they look distasteful.

圖 / 李世仰（Photo：Shih-Yang Lee）

含毒性物質不好吃的赤星瓢蟲。
Ladybird, *Lemnia saucia*, is toxic and unpalatable.

常見的瓢蟲體會分泌有毒或刺激性物質，外形代表不好吃；無毒的金花蟲及蜚蠊也演化出紅斑外形，欺騙捕食者我也不好吃。

Ladybirds can secrete toxic or irritating substances, which keeping predators away ; some non-toxic leaf beetles and cockroaches are therefore mimic the appearance of ladybird to deceive the predators by announcing that I am not a good meal.

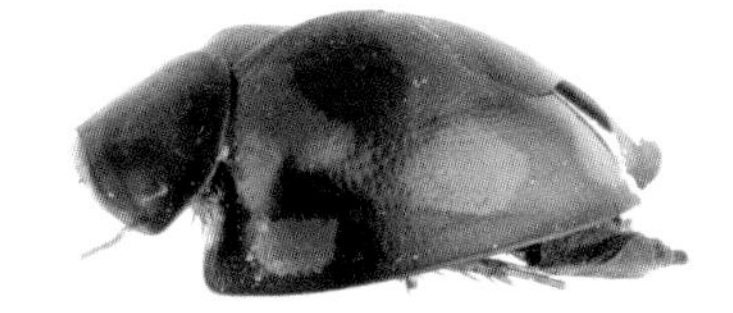

圖 / 李奇峰（Photo：Chi-Feng Lee）
圖 / 許至廷（Photo：Chih-Ting Hsu）

無毒的金花蟲 (*Nonarthra*)（左圖）及擬瓢蠊 (*Prosoplecta*)（右、下圖）模擬瓢蟲外觀，欺騙捕食者我是有毒的。參考自 Lee 2014
The leaf beetles (*Nonarthra*) (Left) and ladybird-mimicking cockroach (*Prosoplecta*) (right & bottom) are nontoxic but they mimic the toxic ladybird to deceive the predators.（Cited from Lee 2014）

特異演化的昆蟲們

Insect, unique product of evolution

黃詩穎、葉文斌、周明勳
Shih-Ying Huang、Wen-Bin Yeh、Ming-Hsun Chou

六、 特異演化的昆蟲們

在環境多變的道路上，不少昆蟲演化出其獨特的存活及傳承方式。

Insect, unique product of evolution

In the changing environment, many insects have evolved their unique ways of survival and inheritance.

圖 / 江允中（Photo：Yun-Chung Chiang）

金黃蜻蜓配對。
Copulation of dragonfly, *Orthetrum glaucum*.

（一）愛的枷鎖

蜻蜓與豆娘雙雙對對，會以為「愛心」交尾姿態就是羅曼蒂克到了極點；那可能就大錯特錯了，這恐怕是強大的生殖競爭壓力所演化出來的生殖策略。雄蟲用特化的尾毛牢牢夾住雌蟲頭後方，使雌蟲腹部末端向前伸至雄蟲第二、三腹節間儲存有精子的交尾器進行交尾，並持續到產卵結束，確保產下的是自己的後代。

Shackling for love

Do you agree the elegant copulation dances of dragonflies or damselflies are so romantic which touch deeply your "heart"? The heart-shaped copulation posture in dragonflies and damselflies is not an intentional act but an obligately evolutionary adaptation in reproduction strategies. The male holds female's pronotum firmly by his terminal abdominal cerci, so that forces the female bends and extends her abdominal terminal genitalia forward to the male's copulatory apparatus located at between the 2nd and 3rd abdominal sternites where his sperm were stored in a reservoir. In some species, the pair will stay in tandem during the whole egg-laying process. These are strategies to make sure the offspring belong to the pair.

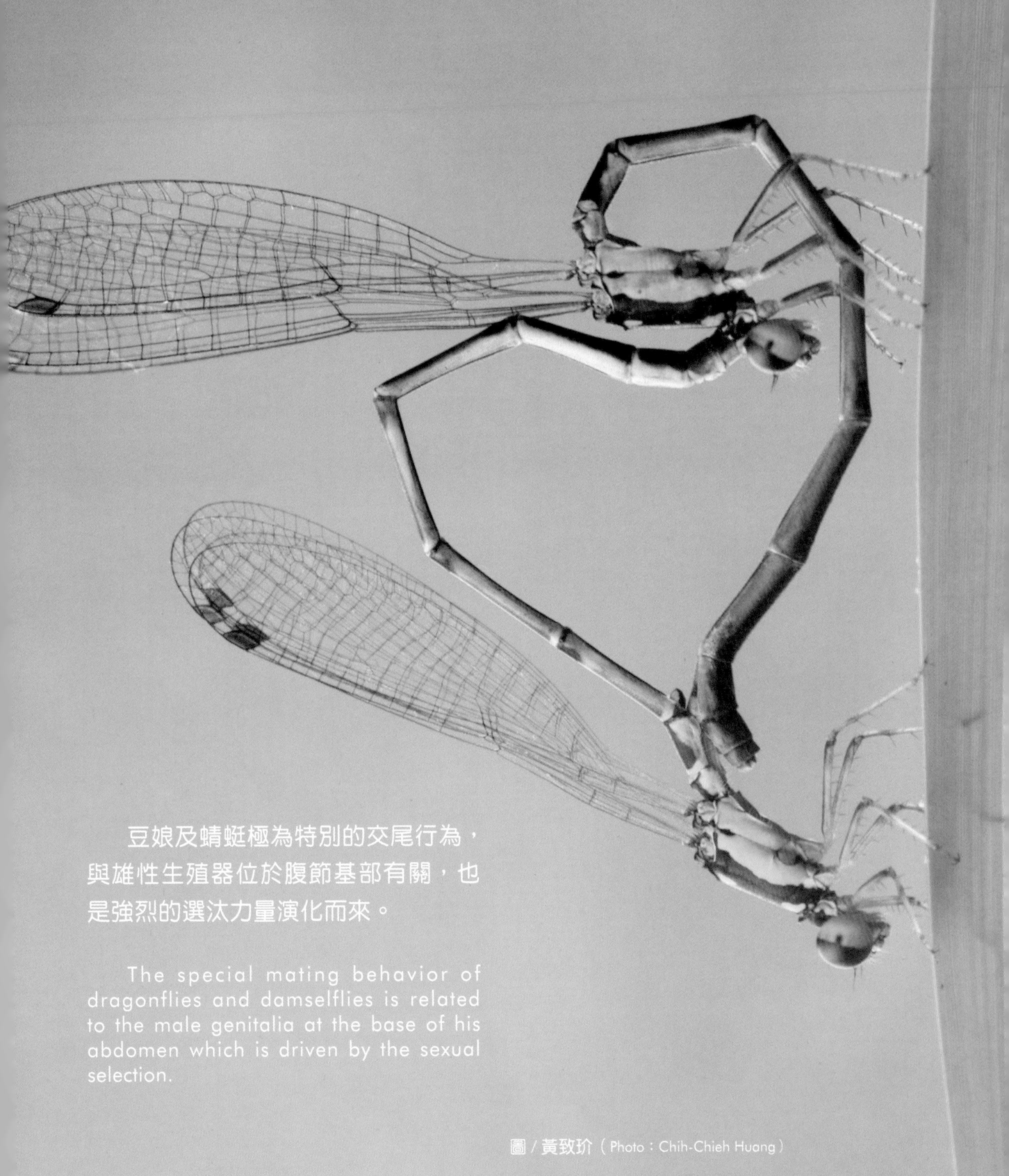

豆娘及蜻蜓極為特別的交尾行為，與雄性生殖器位於腹節基部有關，也是強烈的選汰力量演化而來。

The special mating behavior of dragonflies and damselflies is related to the male genitalia at the base of his abdomen which is driven by the sexual selection.

圖 / 黃致玠（Photo：Chih-Chieh Huang）

四斑細蟌配對。
Copulation of damselfly, *Mortonagrion hirosei.*

(二) 因為愛，我得強壯、再強壯

不像有些小型霧社血斑天牛有「打帶跑群體上」的求偶行為，我們甲蟲的世界就是要「大」才能求得芳心。

I have to be stronger and stronger as a lover

Unlike Wu-she blood-spotted longhorn beetles which perform collaborative courtship behavior in group to outcompete larger competitors, body size matters to ensure successful mating in many beetles.

圖 / 黃致玠（Photo：Chih-Chieh Huang）
圖 / 戴從伊（Photo：Chung-I Tai）

性擇的壓力下，雄鍬形蟲及獨角仙體型都較大。

Sexual selection forces the males of stag beetle, *Odontolabis siva*, and rhinoceros beetle, *Trypoxylus dichotomus*, to have a larger body than females.

圖／廖智安（Photo：Chi-An Liao）
圖／戴從伊（Photo：Chung-I Tai）

長臂金龜雄蟲演化出很長的前足，雌性個體則無。
Males of Formosan long-armed scarab, *Cheirotonus formosanus*, have elongated front legs but not for the female counterparts.

不像多數昆蟲雌蟲個體較大，大型的雄甲蟲在異性競爭上會較優勢。

In insect, females are usually larger than males. However, in beetles, males are larger than females in order to outcompete others for mating.

(三) 擋不住的誘惑：情味傳千里

昆蟲觸角的形狀多樣特異，主為觸覺及嗅覺的功能；不少蛾類演化出複雜的羽狀觸角，在烏漆抹黑中，得以接收到愛蟲的性費洛蒙，繁延種族。

An irresistible allure: Long-distance spreading odor

Antennae of insects have well-developed olfactory and tactile senses. Many moths have evolved complicated bipectinate antennae which allow them to receive sex pheromone from their partner in darkness.

一些雄蛾發達的雙櫛齒狀觸角，是有效接收性費洛蒙的特化構造。

Some male moths have a pair of bipectinate antennae which allow them effectively to receive sex pheromones.

圖 / 戴從伊（Photo：Chung-I Tai）

樹形尺蛾有發達的觸角。
The geometrid moth, *Erebomorpha fulguraria*, has a pair of well-developed bipectinate antennae.

圖 / 江允中（Photo：Yun-Chung Chiang）

姬長尾水青蛾有發達觸角。
The moon moth, *Actias neidhoederi*, has a pair of well-developed antennae.

(四) 腹背受敵

美麗的蝴蝶並非與天俱來，而是演化雕琢而成；亮麗的翅背常與性擇 (sexual selection) 有關，而翅腹則是性命攸關的天擇 (natural selection)。

Have attacks on dorsal and ventral

The beauty of butterflies is not a free gift, but a consequence of the succession in evolutional processes. The gorgeous dorsal wing patterns are usually associated with sexual selection, while the ventral wing patterns are formed through natural selection.

圖 / 戴從伊 (Photo : Chung-I Tai)
圖 / 詹明澍 (Photo : Ming-Shu Chan)
圖 / 江允中 (Photo : Yun-Chung Chiang)

枯葉蝶翅背翅腹及翅背。
Dorsal and ventral view of the dead leaf butterfly, *Kallima inachus formosana*.

圖 / 廖智安 (Photo : Chi-An Liao)
圖 / 詹明澍 (Photo : Ming-Shu Chan)

孔雀青蛺蝶翅背及翅腹。
Dorsal and ventral view of the peacock butterfly, *Junonia orithya*.

孔雀青蛺蝶及枯葉蝶的亮麗翅背可吸引異性，暗色的腹面有助於隱藏躲避敵人。

The bright color on dorsal wing of the peacock butterfly and the dead leaf butterfly can help them to attract their counterpart. The dark color on ventral wing helps them to hide from the enemy.

(五) 我很臭！可是我超有母愛

椿象這類昆蟲臭雖臭，卻演化出各類的護幼行為；有些會趴在卵塊上保護、有些會直接背卵趴趴走、有些會守護於周圍隨時照顧，這一切都是為了下一代的存活。

I'm stinky, but I do all the jobs that the best mother could do

True bugs have foul-smelling odor, yet they evolved different types of parental care to protect their eggs and young. Adults may carry the eggs on the back all the time until the eggs hatched, while the other females stay closely to guard the eggs. All these behaviors are likely to sure their offspring survival.

圖 / 黃致玠（Photo：Chih-Chieh Huang）
圖 / 李世仰（Photo：Shih-Yang Lee）

黃盾背椿象護卵及及黃斑椿象群聚若蟲。
Parental care of the jewel bug, *Cantao ocellatus*, and the grouping nymphs of *Erthesina fullo*.

　　俗稱臭蟲的椿象是不受喜愛的昆蟲，牠們卻演化出昆蟲少有的護幼行為。

　　Stink bugs, commonly known as true bugs, have evolved the unique behavior of parental care to protect their eggs and offspring.

圖 / 廖智安（Photo：Chi-An Liao）
圖 / 林冠妤（Photo：Kuan-Yu Lin）

大田鱉抱卵的照護行為。
Parental care of the giant water bug, *Kirkaldyia deyrollei*, guards the egg mass.

圖 / 廖智安（Photo：Chi-An Liao）
圖 / 黃福盛（Photo：Dash Huang）

負子蟲背卵。
Water bug, *Diplonychus esakii*, carries and guards the eggs on its back.

圖 / 廖智安（Photo：Chi-An Liao）

橙斑埋葬蟲製作的肉團及取食的幼蟲。
Carrion beetle, *Nicrophorus nepalensis*, makes meat ball for its larvae.

(六) 替寶寶做好育嬰房

埋葬蟲將動物屍體加工做出一顆顆肉丸子，產卵其中，確保幼蟲寶寶有得吃。糞金龜用後足將糞球推到巢室中，再產入一卵，幼蟲於內取食糞便成長化蛹；雖臭，但吃住不愁！

Set up nurseries well for baby

A carrion beetle makes dead animal debris into balls and lays its eggs therein to ensure the young has plentiful of food supply. The dung beetle uses its hind legs to roll the dung-ball into her nursery and then lays one egg in it. Her baby feeds on dung and grows in the dung-ball. The balls look disgusting but a perfect nursery for the beetles.

埋葬蟲及糞金龜不僅確保下一代有吃住不愁的育嬰室，還會將牠們藏於地下的保護窩。

Burying beetles and dung beetles provide not only nursery and food but also a protected shelter for young to live.

圖 / 李世仰（Photo：Shih-Yang Lee）
圖 / 黃致玠（Photo：Chih-Chieh Huang）

糞金龜及其製作的糞球。
Dung beetle, *Paragymnopleurus* sp., and its dung ball.

（七）超人氣特異昆蟲

昆蟲常有奇奇怪怪的形狀，令人驚奇喜愛。如不是鞘翅目的甲蟲，卻演化出硬翅的盾背椿、瓢蠟蟬、鎧蠅等及其它特化的附器及構造的昆蟲。

Extraordinary insects, super-popular

We may sometimes find the strange and lovely looking insects, like specialized appendages and structures. For example, jewel bug, issid planthopper, and beetle fly etc. do not belong to the coleopteran beetle, yet they have hard elytra.

圖 / 黃宸瑋（Photo：Chen-Wei Huang）

瓢蠟蟬
Issid planthopper, *Gnezdilovius* sp.

圖 / 唐昌迪（Photo：Chang-Ti Tang）

鎧蠅
Beetle fly

圖 / 黃宸瑋（Photo：Chen-Wei Huang）

犨蹙渫蜣
Dung beetle, *Onthophagus rugulosus*

圖 / 雷喬安（Photo：Chaou-An Rei）

食材性拉氏木蠊
Wood feeding cockroach, *Salganea raggei*

圖 / 李世仰（Photo：Shih-Yang Lee）

四斑柄眼蠅
Stalked-eye fly, *Teleopsis quadriguttata*

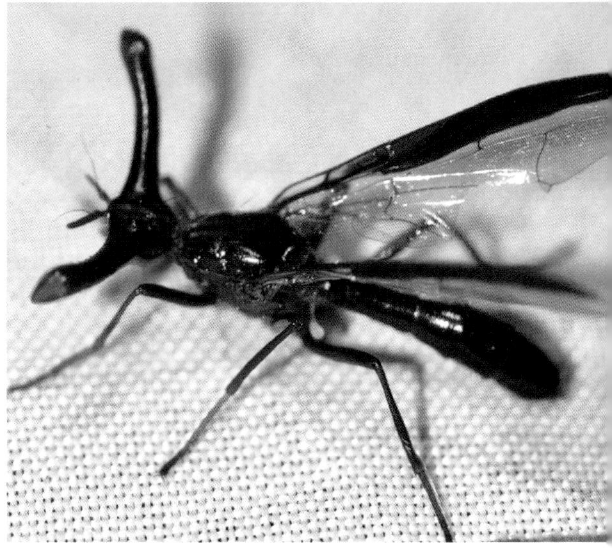

圖 / 蔡正隆（Photo：Cheng-Lung Tsai）

黑眼擬柄眼果實蠅
The stalked-eye fruit fly, *Pseudopelmatops continentalis*

圖 / 廖智安（Photo：Chi-An Liao）

台灣長臂金龜
Formosan long-armed scarab, *Cheirotonus formosanus*

圖 / 蔡正隆（Photo：Cheng-Lung Tsai）

渡邊氏東方蠟蟬
Long rostrum planthopper, *Pyrops watanabei*

圖 / 唐昌迪（Photo：Chang-Ti Tang）

紹德錨角蟬
Treehopper, *Leptobelus sauteri*

演化出奇特造型的昆蟲是牠們受大眾喜愛的重要原因之一；如像甲蟲的鎧蠅、圓飛蝨及深山蜚蠊，頭有突出物的犟蹩渫蜣及長吻白蠟蟬，眼睛長在長柄上的柄眼蠅及果實蠅，有特化構造的長臂金龜及角蟬。

Insect evolved fascinating and extraordinary structures, which enchant us and make them so popular. For examples, the beetle-like beetle fly, issid planthopper, and wooden roach; the protruding head of dung beetle and Watanabei long rostrum planthopper; the protruding eyes of stalk-eyed fly and stalk-eyed fruit fly; and the elongated structure of the Formosan long-armed scarab and treehopper.

(八) 如夢幻妙影

埋物理色又稱為構造色，光照射到昆蟲表皮時，因其突起不規則排列，干涉、折射後讓光反射出如夢似幻般的燦爛顏色。

Insects embody fantastic kaleidoscope

Physical or structural coloration of insect is the production of colour through interference, refraction, and mirrored light by microscopically irregular arrangement of structured surfaces.

圖 / 鄒議慷（Photo：Yi-Kang Tzou）

白邊魔目夜蛾翅鱗片呈現的構造色。
Structural color on the wing scales of *Erebus albicincta obscurata*.

翅膀不同角度會閃爍出耀眼光芒，是有些昆蟲驅敵避凶的法寶，短暫片刻已讓牠們足以逃離被捕的命運。

Many insects use light-interacting structures on their wing scales to produce brilliant iridescent colors. The short period of time of manipulating light results them to escape from predators.

圖 / 李世仰（Photo：Shih-Yang Lee）
圖 / 詹明澍（Photo：Ming-Shu Chan）

圓翅紫斑蝶及小紫斑蝶幻麗的物理色。
The shining physical color of purple butterflies, *Euploea eunice* and *Euploea tulliolus koxinga*.

圖 / 鄒議慷（Photo：Yi-Kang Tzou）

大彎爪蟻蛉翅膀呈現的構造色。
Structural color on the wing scales of the *Thaumatoleon splendidus*.

（九）高山孤島的遺傳獨特性

高聳入雲的山頭如孤島般，限制了適應於高山的生物族群間的基因交流(gene flow)，歷經數十萬或百萬年，走入異域種化 (allopatric speciation) 的路途與多樣的遺傳組成。

Mountain-island isolation & Genetic differentiation

Mountain towering into the clouds resembles isolated islands that inhibited gene flow between isolated adaptive alpine populations. Allopatric speciation and diverse genetic compositions are therefore gradually formed over tens of thousands or millions of years.

繪 / 朱能榮（Paint：Neng-Jung Chu）

高海拔適應的昆蟲僅能存活於高山，多已被隔離在各山頭。參考自 Weng et al. 2016
Alpine insects can only survive at high altitude. Most of them are isolated by mountains.（Cited from Weng et al. 2016）

演化適應的瓶頸

The bottleneck of evolutionary adaptation

李威樺、葉文斌、楊曼妙
Wei-Hua Li、Wen-Bin Yeh、Man-Miao Yang

七、 演化適應的瓶頸

十年河東、十年河西；在千百萬年的演化路上，生物也會面臨種族生死存亡的窘境，大家所熟悉的保育類昆蟲多為此。

The bottleneck of evolutionary adaptation

The world is changing unexpectedly. Over millions of years of evolutionary history, organisms might meet the state of extinction. Those in conservation would be facing their survivorship.

繪 / 朱能榮（Paint：Neng-Jung Chu）

被命名即已絕種的楊氏淺色小豹蛺蝶。參考自 Hsu and Yen 1997
By the time Yang's light-colored leopard butterfly was named, it has become extinct.（Cited from Hsu and Yen ）

(一)保育昆蟲需要先經過資格考?!

地理位置特殊、環境多樣特異的台灣島,經歷更新世 (Pleistocene) 多次冰河洗禮,生物多樣性高無用置疑。

在時代的洪流下,拚經濟的土地開發利用已停不下來,結果就是這些野生動物面臨存活的壓力。若原本已是區域性分佈、食性專一、遷移力弱、族群小或僅住於高海拔,能存活都已不易了,如再加上大型、亮麗易遭受獵捕販售或棲地破碎化,命運可想而知。在台灣,面臨生死存亡需要保育的昆蟲太多了,僅能選一些大家較有感的物種來代表蟲蟲們說說話!

並不是大型亮麗就可以成為保育類昆蟲,得先有相關的生態資料,經由專家們一連串的審核評估,再經過分數等級,才有可能成為保育類昆蟲,分數太低,還不及格呢?

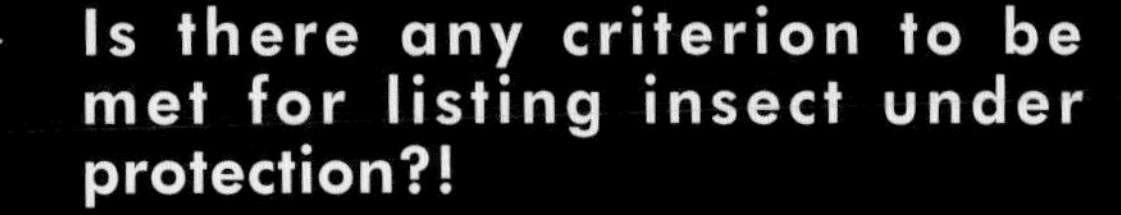

Is there any criterion to be met for listing insect under protection?!

Taiwan experienced glacial cycles during the Pleistocene. Its distinctive geographic location and environmental diversity leads her the great biodiversity.

Land exploitation for economic activity have become an irresistible trend and making survival an overwhelming challenge to wildlife. Due to local distribution, specialized food habit, low capacity in migration, small population, or being restricted to high altitude areas, some insects struggle for existence. Hunting, wildlife trade, and habitats fragmentation further aggravate the situation in Taiwan, there are numerous insects are endangered and need to be protected. The book compiles a list of a few protected insect species

The species need protected is not simply solely depended on its size of body or the gorgeous appearance. To be listed as protected insect, a series of evaluation based on pertinent data is necessary. Only those categorized in endangered level are protected. Will you say this is a qualifying exam for protected insect species.

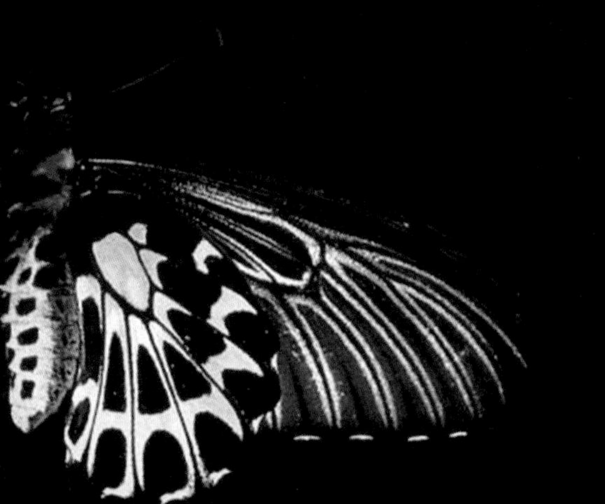

圖/吳怡欣(Photo:I-Hsin Wu)

棲地破壞是黃裳鳳蝶受保育的主要原因。
Habitat destruction is the main issue when it comes to insect conservation of the golden birdwing butterfly, *Troides aeacus formosanus*.

繪 / 朱能榮（Paint：Neng-Jung Chu）

棲地破壞是保育類昆蟲面臨的最大問題。
Habitat destruction is the first issue when it comes to insect conservation.

圖 / 李惠永（Photo：Hui-Yong Li）

寬尾鳳蝶的食草稀少且食性專一，是其受保育的主要原因。
Because of the specific host plant of the broad-tailed swallowtail butterfly, *Agehana maraho*, is rare, as it has been listed as one of protected species.

圖 / 楊曼妙（Photo：Man-Miao Yang）

隔離在棲地狹小島嶼的球背象鼻蟲是其受保育的原因。
Because of the limited population size and distributed in a small isolated island, *Pachyrrhynchus* weevils are highly protected.

圖 / 吳怡欣（Photo：I-Hsin Wu）

棲地範圍小且易受破壞是津田氏大頭竹節蟲受保育原因。
The tsudai big-headed stick insect, *Megacrania tsudai*, is one of the protected species because its habitat is limited and easily to be vulnerable.

黃裳鳳蝶
The golden birdwing, *Troides aeacus formosana*

(二)台灣的保育昆蟲 Protected insect in Taiwan

1. 挑食的金披風 — 黃裳鳳蝶 (*Troides aeacus formosana*)

說到後翅金黃亮麗的黃裳鳳蝶，通常會聯想到風光明媚的恆春半島，但台灣其他地區仍可偶見其芳蹤。身為鳥翼蝶的一員，來去如風，幼蟲專食「馬兜鈴」的有毒植物，這是牠防禦天敵的一種方法。因大型美麗、食性專一及獵捕壓力曾讓牠的族群非常有限。在棲地復育及賞蝶活動的保育推廣下，黃裳鳳蝶族群量已增多，行政院農委會已自二級調為三級保育類。

A picky eater in a golden wing – The golden birdwing, *Troides aeacus formosana*

Troides aeacus formosana is a butterfly with golden and gorgeous hind wings that mainly distributed in the beautiful scenery Hengchun Peninsula. In fact, this species is found throughout Taiwan island. It is a member of bird-wings butterflies which can fly swiftly like a gust of wind. The larva eats only the poisonous *Aristolochia* plants by which help them carry the toxic substances to protect themselves from predation. Its population used to be very small because of its large-size, specific diet and beautiful appearance make them a main target of insect collector. Fortunately, as a result of conservation effect on habitat restoration and butterfly ecotourism, its population has increased. Therefore, Council of Agriculture, Executive Yuan has adjusted the level of this protected species from Class II to Class III.

2. 海岸林投內的"香竹" — 津田氏大頭竹節蟲 (*Megacrania tsudai*)

我是可長到 12 公分的大型竹節蟲，僅住在台灣東南部及綠島蘭嶼等黑潮流過海岸邊的林投上，叫做「津田氏大頭竹節蟲」，行孤雌生殖 (parthenogenesis)，遇到危險時，會自前胸腺噴出刺激性薄荷味的防禦性物質給你好看。因體型大、棲地小、僅吃林投葉、獵捕壓力又大，遂成了台灣珍貴稀少的昆蟲之一。其實，我並不是仰慕墾丁迷人的風光，而是隨黑潮漂啊漂～漂來的。最近有研究顯示，我們的遺傳變異很少；而且，我也吃山林投的，這麼吃著吃著，就吃出雄性個體來了！請讓我住久一點吧！台灣還不錯住！(參考自 Wu et al. 2020)

Bamboo sticks insect on coastal *Pandanus ordoratissimu* – Tsudai big-head stick insect, *Megacrania tsudai*

I, *Megacrania tsudai*, is a large stick insect with body size up to 12 centimeter, reproduce by a way of parthenogenesis. I feed only on *Pandanus ordoratissimu* and distribute over the southeastern Taiwan and offshore isles of Green and Orchid where the Kuroshio Current flows by. When I am disturbed, I will spray irritant, a mint-smell defensive substance from prothoracic gland. The large size with limited habitats as well as eating specifically on *Pandanus* leaves and being confronted hunting pressure, the bamboo stick is always vulnerable in Taiwan. To be honest, I — the bamboo stick living at Kenting is not for its beautiful landscape, but flew there by the Kuroshio Current. Recent research shows I am in low genetic variation. By the way, I also eat *Freycinetia formosana*, that might be why the male individuals have been forced induction. Please let me stay here as long as possible, if you stop to hunt me down. Taiwan is a pretty nice place to live! (Cited from Wu et al. 2020)

1 | 2 繪 / 李家淇 (Paint : Jia-Qi Li)

3. 讓分類學者踢到大鐵板的虹彩叩頭蟲 (*Campsosternus watanabei*)

夏季裡的中海拔山區，偶會見到一道美麗的彩虹掠過林間，那不是彩虹，而是艷麗的「虹彩叩頭蟲」，屬二級保育類。關於虹彩叩頭蟲的命名，有著一個讓昆蟲分類學者吃足了苦頭的美麗錯誤，甚至有公告保育類圖片並非「真身」的爭議。牠最早曾被認為中國朱肩麗叩頭蟲 (*C. gemma*) 的一型色或亞種，也曾被命為紅緣大青叩頭蟲 (*C. yasukii*)，後來終於改名為虹彩叩頭蟲 (*C. watanabei*)，但爭議仍在；直到 2014 年應用大量的樣本及 DNA 資料，才有了正名，也解決保育對象失真的困擾，確定這道美麗的彩虹是真的存在於台灣的，約 150 萬年前與同樣美麗的朱肩麗叩頭蟲分化開來。(參考自 Hsieh et al. 2014)

A snag to taxonomists, click beetle *Campsosternus watanabei*

In summer days, if you trolled at mid-elevation mountains in Taiwan may find the rainbow sheath click beetle, *Campsosternus watanabei*, flitting. It is protected under a category II protected species. Behind its name, there is a story remarking a peculiar mistake that really troubled many insect taxonomists. The identity has confused scientists for decades, even scientists argued the photo of the click beetle shown in the Protected Species Catalogue. It was initially recognized as a subspecies of *C. gemma* and has been mistakenly erected as *C. yasukii*. The beetle was renamed as *C. watanabei*, but the debates remained. It is not until 2014, a detail job of comparison and DNA analyses from a large number of specimens, the name was finally proven, corrected and verified, the rainbow indeed exist in Taiwan and was differentiated from its beautiful *C. gemma* sister about 1.5 million years ago. (Cited from Hsieh et al. 2014)

1 | 2 繪 / 李家淇（Paint：Jia-Qi Li）

4. 櫻花樹上的紅寶石 — 霧社血斑天牛 (*Aeolesthes oenochrous*)

風光明媚，正是春天賞櫻好季節，你可能會發現樹幹下方有著木屑堆積，那就是這個保育類昆蟲－霧社血斑天牛幼蟲取食造成的。大型艷麗的霧社血斑天牛，曾是大家競相收藏的天牛寶貝之一。因性擇 (sexual selection) 的擇偶行為，大型雄蟲較有優勢，可大到 8 公分，但打不贏的小個兒雄蟲們，則以打帶跑趁機上的群體戰術，保有了繁衍子代的機會。感謝賞櫻旅遊活動，牠的子孫變多了，保育指數終於降低了；不過！不過！千萬不要把牠當害蟲消滅，因為大型豔麗的我獵捕壓力一定還會非常非常大，美麗總不能說是個錯誤吧！(參考自 Wei 2009)

Rubies on cherry trees – Wu-she blood-spotted longhorn beetle, *Aeolesthes oenochrous*

The spring sunny day is a good tourism for cheery blossom. You may see wood powder piling up at the bottom of cherry tree, which is produced by larvae of *Aeolesthes oenochrous* when feeding. This protected longhorn beetle species is large in size with beautiful looking is used to be a highly wanted insect by collectors. According to the sexual selection, a larger body is at an advantage and the male may grow up to 8 centimeter. However, if the small males fail in competition, they strive for their reproductive opportunity by using a teamwork strategy. Owing to the activities of cherry blossom viewing with much attention has been planted on cherry tree, the population of the beetle started to increase and the degree of protection urgency has finally been slightly lowered. However, please do not kill them like pest, because its large and gorgeous, hunting pressure could by no means mitigated. Come on, don't blame for my beautiful looks! (Cited from Wei 2009)

5. 我很紅，紅進紅皮書了！國寶蝶寬尾鳳蝶 (*Agehana maraho*)

我是被列為 IUCN 紅皮書保育的寬尾鳳蝶，於日治時期即已被列為天然紀念物，目前大多活在北部中海拔山區。說起我的幾十種遠祖親戚們可都是「留美」的呢！雖然我與中國寬尾鳳蝶的共同祖先早於 1800 萬年前即已經由白令陸橋來到了東亞，但我則是約 10 萬年前才由海峽陸橋來到台灣的新住民，也僅剩檫樹可吃。千萬別再抓我了，連我吃的檫樹都受到保護了，更別說是我；請多多加緊保護我這個天然紀念物，別讓我真的成為紀念物。(參考自 Wu et al. 2015)

The national treasure of the broad-tail swallowtail butterfly, *Agehana maraho* in Red List

Agehana maraho was listed in the IUCN Red List and had been recognized as one of natural monuments since the Japanese occupation years of Taiwan. Most of them live in the middle-altitude mountains of northern Taiwan. You may know, tens of spices of my ancestral relatives are US-based. While the common ancestor of *Agehana elwesi* and I came to East Asia by way of the Beringia land-bridge at as early as 18 million years ago. I was a new immigrant at one hundred thousand years ago and fed only on sassafras. Please don't try to catch me anymore. Even my food, sassafras, is protected, let me alone, will you? Please protect me as a natural monument and don't let me become a memorial species. (Cited from Wu et al. 2015)

繪 / 李家淇 (Paint：Jia-Qi Li)

6. 在離島遇見幸福？如珠寶般亮麗的球背象鼻蟲 (*Pachyrhynchus* spp. and *Eupyrgops waltonianus*)

要看牠們請來蘭嶼及綠島，據傳島嶼居民用捏碎堅硬翅鞘以彰顯自己強壯，譬喻可讓愛人幸福。仔細看，你會發現近完美圓形、如珠寶般的堅硬球背共有6種，是被大量捕捉販賣的主因。菲律賓系的球背象鼻蟲在台灣是看不到的，因此讓當年的博物學者鹿野忠雄驚為天人，有將生物地理分布區隔的華萊士線北延到台灣及蘭嶼之間的想法。島民藉幸福蟲見證幸福，那「幸福蟲」自己呢？堅硬外殼的確使吃牠們的蜥蜴寧願吐出來，閃亮顏色也宣告著不好吃的警戒意味。因被捕壓力大、棲地小、食物有限及土地開發，數量急遽減少；有些珠寶雖不難得見，但因六種外型相像，恐難區分，遂同列珍貴稀有二級保育類昆蟲，藉由整體保育的概念來達到保育目的。(参考自 Wang et al. 2018)

Have you "met happiness" on off-shore islands? The sparkling jewelry easter egg weevils, *Pachyrhynchus* spp. and *Eupyrgops waltonianus*

If you want to see the easter egg weevils, you should go to the Orchid and Green Isles. According to a legend, if a resident can crush the bug's rigid elytra by hand, this means he is strong and capable to bring happiness to his lover. If you observe the beetle closely, you'll find six patterns on their elytra. Their dome-shaped elytra is so rigid as a ruby. This is why they are always illegally collected and traded. Those *Pachyrhynchus* & *Eupyrgops waltonianus* beetles found in Philippine are absent from the Island of Taiwan. Based on the finding, Tadao Kano, a naturalist suggested to extend the Wallace's Line northward between Taiwan and Orchid Isle. Residents believe that the beetle may bring luck to them if they see the beetle. The hard exoskeleton makes them unpalatable for the predatory lizards but have to spit its prey out. The shiny color of the beetle gives sign to predator that they are distasteful. Nevertheless, under hunting pressure, limited habitats and food supply, and excessive land development, the population of the snout beetle is decreasing sharply. There are six easter egg weevils are similar in appearance and are not easy to differentiate them. Although some are not rare, the six snout beetles have been all listed as Class II protected insects. (Cited from Wang et al. 2018)

繪 / 李家淇（Paint：Jia-Qi Li）

球背象鼻蟲
The easter egg weevils, *Pachyrrhynchus* spp. and *Kashotomus multipunctatus*

繪 / 李家淇（Paint：Jia-Qi Li） 1｜2

7. 甲蟲的「黑金」 — 長角大鍬形蟲 (*Dorcus schenklingi*) 及台灣大鍬形蟲 (*D. hopei formosanus*)

廣受歡迎的鍬形蟲當中，這兩種保育類外型並非特別美麗，只因擁有巨大的體型、雄壯的大顎而深受喜愛，也如此遭受非法獵捕、販售，再加上棲地開發破碎化，一直都處於需要保育的狀態，尤其是長角大鍬，列為珍貴稀有的二級保育類。因性擇 (sexual selection) 的關係，生性兇猛如黑金剛的雄蟲可長成 9 公分的巨無霸，大顎長、如彎月闊刀；全台中、低海拔廣泛分布，喜青剛櫟、栓皮櫟等殼斗科植物汁液，常匿於樹洞內。至於台灣大鍬則是個分類鐵板，自 1929 Miwa 發表以來，學名多所更迭，目前被視為中國大鍬形蟲下的一個亞種。

"Black gold"in beetles – Formosan long-fanged stag beetle, *Dorcus schenklingi* and Formosan giant stag beetle, *D. hopei formosanus*

Among the popular stag beetles, these two species are not the most beautiful ones, they are eventually suffered illegal hunting and traded continuously in market due to their big body size and magnificent mandibles. The increasing habitat fragmentation and over land reclamation made them in protection list. *Dorcus schenklingi* is recognized as being rare and vulnerable is listed as a Class II protected species. Results from processes of sexual selection, the male is so ferocious as gorillas has a size up to 9 cm with long and tulwar-like mandibles. This beetle is widely distributing in the low- to middle-altitude forests. The beetle likes juicy Fagaceae plants, such as ring-cupped oak and oriental oak, and often hides in the tree holes. As to identify *D. hopei formosanus*, it's a troublesome job to taxonomists. Miwa described it in 1929, its scientific name had been changed alternatively, and was currently recognized as a subspecies of *D. hopei*.

8. 蜻蜓界的巨無霸 — 無霸勾蜓 (*Anotogaster sieboldii*)

聽我的名字就知道，我這個巨無霸是台灣最大的蜻蜓，可長到十一公分；雖非台灣特有，但外型已與日、韓、遠東、中國族群有別，據說已有亞種的架式；台灣雖是我家鄉的最南方，但北台灣的乾淨水域較容易找到我。近年工業發展及土地開發，我被迫搬到較少人住的山裡，族群量雖然還不少，但因特殊的地理分佈南界及巨無霸的大明星的氣勢，頗受歡迎；希望繼續保育我，不要讓我在台灣沒得住！

A giant in dragonfly – Jumbo dragonfly, *Anotogaster sieboldii*

It is just like my name implies, a giant. I am the largest dragonfly in Taiwan, with a body length up to 11 cm. Even I'm not endemic to Taiwan, my features actually distinctive from the close relatives in Japan, Korea, the Far East and China. And thus a new subspecies was suggested for me. Taiwan is the south limit of my geological distribution and you can often see me more around a clean stream in northern Taiwan. Recently, due to industrial development and land exploitation, I was forced moving into desolate mountains. The special geological distribution of southern boundary and majestic giant size make pathetical me to be hunted. Although my population is not low, I need your cares for protection. Please don't make me vanished from Taiwan!

9. 黑色樂師 — 台灣爺蟬 (*Formotosena seebohmi*)

老婆不嘮叨的蟬總是用牠響亮的叫聲告訴我們夏天到了，中南部的低海拔森林中，更有一身穿黑袍、聲音嘹亮的聲樂家一「台灣爺蟬」，是台灣最大型的蟬類，翅展可達 15 公分；胸背不僅有著迷人的翡翠條紋，還帶著可愛的面具呢？因大型、亮麗又可愛而廣受歡迎，族群量一直多不起來，雖已列入保育，但低海拔原始森林的開發破壞，更是雪上加霜。若再這麼破壞下去，以後要聽到牠的嗓音恐怕有困難了！

Black musician – Formosan giant cicada, *Formotosena seebohmi*

Every year, when cicada, having unnagging wife, begins to sing, this indicates the beginning of summer. In low-altitude forest of the middle and southern parts of Taiwan, there is a special cicada, seemingly a group of vocalists singing loudly in black color suit of *Formotosena seebohmi*. This species is the largest among cicadas in Taiwan, with a wingspan up to 15 cm. Its notum is covered by a fascinating emerald stripes and they look like a mask. Because of large, gorgeous and adorable feature, this insect is widely hunted for collection and the population growth becomes stagnant. Although it is on the protected list, with the increasing the primary low-land forest destructions, we may one day no longer hear the bugle calls in Taiwan.

10. 偽裝達人 — 蘭嶼大葉螽蟴 (*Phyllophorina kotoshoensis*)

具隱身術的螽蟴科昆蟲，隱蔽於環境不易觀察是出了名的，偽裝達人「蘭嶼大葉螽蟴」更是箇中翹楚，菱形狀的胸背，側看就像一片葉子，令人難以發現。偏愛山葡萄，但火筒樹、大戟科、茶茱萸等植物也吃。因侷限於蘭嶼、寄主偏好、喜森林邊緣，加上近年觀光興盛，大量整治道路、橋樑、野溪與海岸林，棲地嚴重被破壞，讓數量原已不多的牠們被歸類為台灣珍貴稀有的二級保育類動物。

The master of camouflage – Lansu giant katydid *Phyllophorina kotoshoensis*

Insect family Tettigoniidae is famous for its masquerade and camouflage to its background. Among the katydids, *Phyllophorina kotoshoensis* is the best one in resembling the surrounding vegetation. Its rhombus protergum looks just like a leaf from both sides rendering it a perfect mimesis. It has a specific host preference. While *Vitis amurensis* is its favorite food, it also consumes Manila Leea and some Euphorbiaceae and Icacinaceae. The katydid is only distributing in Lansu Isle and commonly found at forest edge. As tourism activities blooming in the island, streams and forests were disturbed to make way for road and bridge construction. Couple with their small population, human activities has drastically decreased their population. It is now classified as Class II protected species in Taiwan.

繪 / 李家淇（Paint : Jia-Qi Li） 1 | 2

適得其所的地球村

Find your niche on the earth

周明勳、葉文斌
Ming-Hsun Chou、Wen-Bin Yeh

八、 適得其所的地球村

演化的道路上，硬碰硬的競爭選汰並非都是生物的最愛；為減少生存壓力，也增加自身存活繁衍的機會，生物演化出獨特性以降低競爭，生活於其所喜愛的棲所。但是，人類這個物種對地球的開發已停不下來，如何減緩破壞，增加棲所的多樣性，已是地球村最重要的議題。

Find your niche on the earth

On the evolutionary process, intense competitions for survival are not always favored by organisms. In order to reduce survival pressure and increase life span, organisms evolved peculiar to reduce the degree of competition pressure to occupy a favorable habitat. However, today, exploitation of natural resources from human activities is unstoppable. How to slow down the damage and to diversify the habitats are the first of all jobs for us.

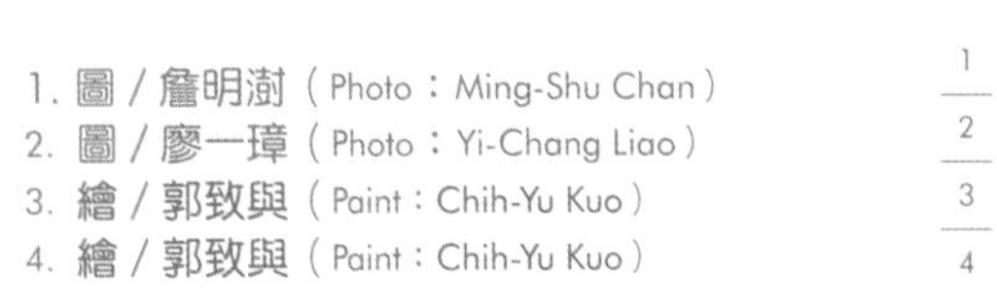

1. 圖 / 詹明澍（Photo：Ming-Shu Chan）
2. 圖 / 廖一璋（Photo：Yi-Chang Liao）
3. 繪 / 郭致與（Paint：Chih-Yu Kuo）
4. 繪 / 郭致與（Paint：Chih-Yu Kuo）

1 / 2 / 3 / 4

水生昆蟲的豆娘水蠆、紅娘華、鼠尾蛆、龍蝨。
Damselfly naiad, , water scorpion, water beetles and rat-tailed maggots are aquatic insects.

圖 / 林峻賢（Photo：Chun-Hsien Lin）
圖 / 江允中（Photo：Yun-Chung Chiang）
圖 / 黃宸瑋（Photo：Chen-Wei Huang）

社會性昆蟲虎頭蜂、螞蟻及白蟻。
Hornet, ants, and termites are social insects.

演化出各式各樣的特性的存活方式，是昆蟲稱霸地球的主因。

Insects evolved various characteristics for survival. This is the main reason making them so successful ecologically.

繪 / 朱能榮（Paint：Neng-Jung Chu）

台灣的中海拔自然棲所有著最高的生物多樣性。
The middle altitude habitat in Taiwan has the highest biodiversity.

(一) 適得其所：生態棲位與棲地的多樣化

在地球演化宿命力的驅動下，硬碰硬的選汰競爭，不如來個最佳演化出路，各自往最適的愛巢及歸宿演化。

Finding your niche: In terms of the diversified niches and habitats

Every organism on the earth is subjected to evolution. You may compete others strongly in the way of natural selection, but evolving to the diversified environment to find the favorable niche may be a more reasonable and feasible way.

(二) 里山倡議

里山 (Satoyama)，是指靠近村落以及周遭山林的地區，里山倡議主張與自然環境和諧共存。淺山保育不但益於生物多樣性，也使人們有安全健康的食物來源與居家環境。

Satoyama Initiative

The term "Satoyama" is referred to area between mountain foothills and arable flat land dominated by human. Satoyama initiative is to seek harmony and coexistence between human and the nature. Conservation effort in low elevation mountains is not only beneficial to overall biodiversity, but also helpful to ensure food security and healthy environment.

繪 / 朱能榮（Paint：Neng-Jung Chu）

與大自然和諧共存的里山倡議，是人類永續生活的重要文化遺產及理念。
The Satoyama initiative to advocate a harmonious life with nature is an important cultural heritage and concept for human to have a sustainable utilization of the earth's resources.

致謝

Acknowledgement

此環境教育推廣書籍的刊印，首要感謝日本學者八田耕吉教授，他於 2010 年捐贈數萬件珍貴昆蟲標本，開啟了中興大學舉辦一系列昆蟲為主題的環境教育推廣活動；這些活動也在國立臺灣科學教育館、國立自然科學博物館、臺北市立動物園等單位的協助下，成功順利展出，並於 2017 年受到李德財校長、呂福興教務長及校方的支持，設立「昆蟲環境教育展示」常設展，接受各學校團體預約導覽推廣環境教育，進而促成了此書的刊印。

本書由 3 位當時的策展人共同編輯策畫，包含八個主題；要感謝昆蟲系師生共同努力、分工完成；更要特別感謝協助文稿撰寫的及提供生態圖片的系友，讓不少簡短的文意藉由圖片得以瞭解。希望本書的內容有助於讀者了解大自然的演化律動，並進而喜愛自然、保護自然。

We thank Kokichi Hatta, the Japanese scholar, who donated tens of thousands of precious insect specimens in 2010, which makes the publication of this book for environment and education promotion possible. We also thank for the assistances and supports from the National Taiwan Science Education Center, National Museum of Natural Science, Taipei Zoo, and thank to the participants to participate in a series of insect exhibition in past years. Furthermore, we thank President Der-Tsai Lee, Dean Fu-Hsing Lu, and the colleagues in campus for their supports to set up a permanent exhibition unit of "Insect Exhibition of Environmental Education" in 2017. The exhibition further promotes the environmental education in natural science to the young generation of school students.

With the supports, we started to compile the exhibition materials into book which containing 8 sections by three editors, i.e. organizers of the exhibition. We wish to express our sincere thanks to the colleagues and students of the Department of Entomology for their efforts in the project. We also thank the alumni of the Department who assisted in the manuscript writing and provided us insect photos which make the book easy to understand by readers at all levels. We hope that you would enjoy reading the book and hopefully, the contents may help you to understand better the evolutional rhythm of the nature, appreciate the nature and protect our mother earth.

參考書目

References

◆ 貢穀紳。1992。普通昆蟲學（中冊）。國立中興大學農院出版委員會出版。

◇ 貢穀紳。1998。普通昆蟲學（上冊）。國立中興大學農院出版委員會出版。

◆ 張書忱。1979。昆蟲形態學。國立編譯館主編。黎明文化事業公司出版。

◇ Futuyma, D J. 2013. Evolution, 3rd ed. Sinauer Associates, Sunderland, Massachusetts.

◆ Grimaldi, D and M Engel. 2005. Evolution of the insects, Cambridge University Press.

◇ Hsieh, J-F, M-L. Jeng, C-H Hsieh, C-C Ko and P-S. Yang. 2014. Phylogenetic diversity and conservation of protected click beetles (*Campsosternus* spp.) in Taiwan: a molecular approach to clarifying species status. Journal of Insect Conservation 18: 1059–1071.

◆ Hsu, Y-F and S-H Yen. 1997. Notes on *Boloria pales yangi*, ssp. nov., a remarkable disjunction in butterfly biogeography (Lepidoptera: Nymphalidae). Journal of Research on the Lepidoptera 34: 142–146.

◇ Lee, -CF. 2014. Review of the genus *Nonarthra* Baly (Coleoptera: Chrysomelidae: Galerucinae: Alticini) from Taiwan and Japan, with descriptions of two new species Japanese Journal of Systematic Entomology, 20: 251–263.

◆ Snodgrass, RE. 1935. Principles of insect morphology, McGraw-Hill Book Company, New York.

◇ Wang, L-Y, W-S Huang, H-C Tang, L-C Huang and C-P Lin. 2018. Too hard to swallow: a secret secondary defence of an aposematic insect. Journal of Experimental Biology 221, jeb172486. doi:10.1242/jeb.172486.

◆ Wei, S-R. 2009. The reproductive behavior of Wushe blood-spotted longhorned beetle, *Aeolesthes oenochrous* (Fairmaire) (Coleoptera: Cerambycidae), and the effects of body size and multiple-mating on reproduction. Master Thesis, Department of Entomology, National Chung Hsing University.

◇ Weng Y-M, M-MYang and W-B Yeh. 2016. A comparative phylogeographic study reveals discordant evolutionary histories of alpine ground beetles (Coleoptera, Carabidae). Ecology and Evolution 6: 2061–2073.doi:10.1002/ece3.2006.

◆ Wu, I-H, H-H Liu, Y-Y Chen, C-L Tsai, Y-C Yu, C-Y Hsiao and W-B Yeh. 2020. Male from geographic parthenogenesis species and life cycles of conservation *Megacrania tsudai* Shiraki (Phasmatodea: Phasmatidae). Entomological Science, online doi: 10.1111/ens.12410.

◇ Wu, L-W, S-H Yen, D-C Lees, C-C Lu, P-S Yang and Y-F Hsu. 2015. Phylogeny and historical biogeography of Asian Pterourus butterflies (Lepidoptera: Papilionidae): A case of intercontinental dispersal from North America to East Asia. PLoS ONE 10: e0140933. doi:10.1371/journal.pone.0140933.

索引 INDEX

國家圖書館出版品預行編目(CIP)資料

昆蟲的華麗變身：演化適應之路 = Insect transformation : an evolutionary adaptation pathway/ 葉文斌，楊曼妙，路光暉編著 . -- 初版 . -- 臺中市：國立中興大學，2022.12 印刷　面；　公分
中英對照

ISBN 978-626-96453-4-3(精裝)

1.CST: 昆蟲學　2.CST: 動物演化

387.7　111019367

昆蟲的華麗變身
演化適應之路

Insect Transformation :
an evolutionary adaptation pathway

編 著 者／葉文斌、楊曼妙、路光暉
責任編輯／郭蕙貞、方光乾
美術設計／斐類設計工作室
發 行 人／薛富盛
總 編 輯／溫志煜
出 版 者／國立中興大學
地　　址：402 臺中市南區興大路 145 號
電　　話：(04) 2284-0291
傳　　真：(02) 2287-3454
服務信箱：press@nchu.edu.tw
出版日期／2022 年 12 月 初版三刷
定　　價／新臺幣 520 元
法律顧問／吳光陸律師

I S B N／978-626-96453-4-3(軟精裝)
G P N／1011101962